SRA

Connecting Math Concepts

LEVEL
BRIDGE

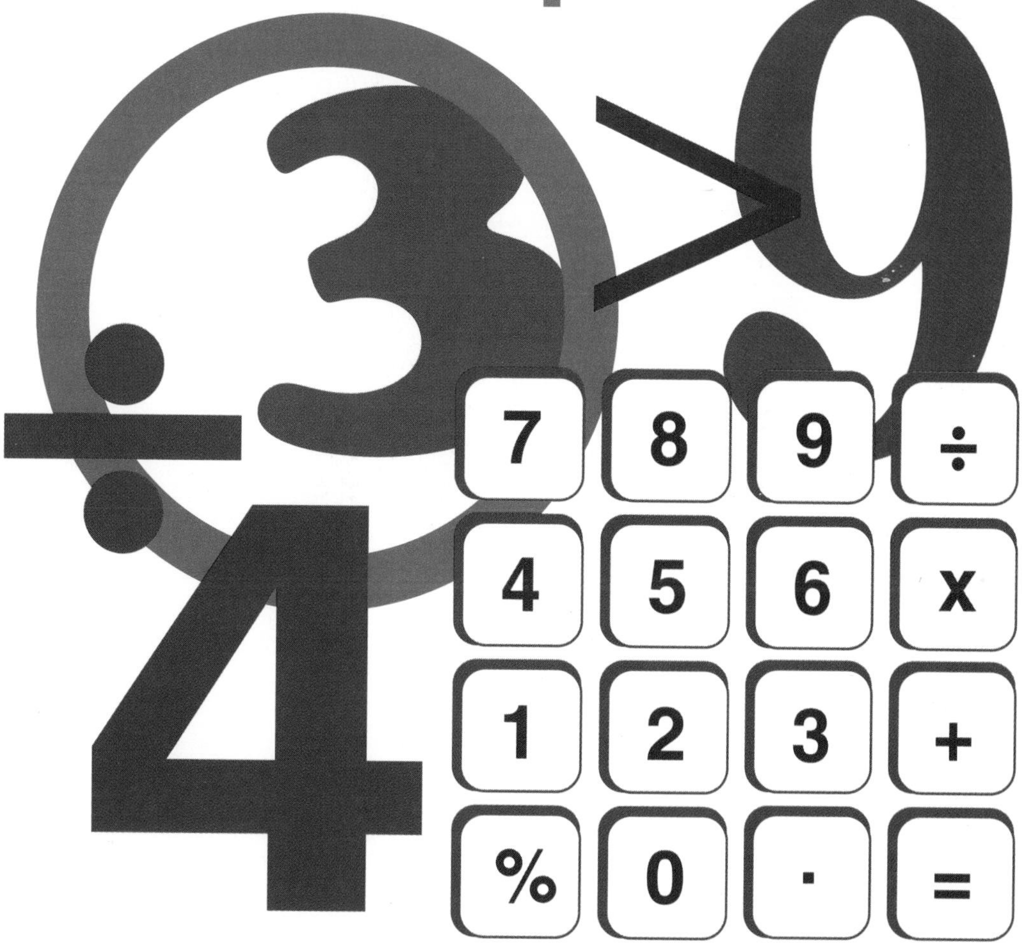

SRA

Columbus, Ohio

The McGraw·Hill Companies

www.sra4kids.com

SRA
McGraw-Hill

Send all inquiries to:
SRA/McGraw-Hill
8787 Orion Place
Columbus, OH 43240-4027

Printed in the United States of America.

ISBN 0-02-684694-2

4 5 6 7 8 9 0 RRC 07 06 05

Bridge
to Connecting Math Concepts

This program will teach you a great deal about doing mathematics. When something new is introduced, your teacher will show you how to work the problems. Later, you'll work the problems on your own, without help.

Remember, everything that is introduced is important. It is knowledge you will need to work difficult problems that will be introduced later.

You must follow your teacher's directions. Your teacher sometimes will direct you to work **part** of a problem, sometimes a **whole** problem, and sometimes a group of problems.

Listen very carefully to the directions. Work quickly and accurately. Most important of all, work hard. You'll be rewarded with math skills that will surprise you.

Lesson 1

Part 1

- Some fractions are more than 1 whole unit. Some fractions are less than 1 whole unit. You can figure out whether a fraction is more than 1 or less than 1 by comparing the top number with the bottom number. The bottom number tells the number of parts in each unit.
- If the top number is **larger** than the bottom number, the fraction is **more than** 1 unit. $\frac{5}{4}$

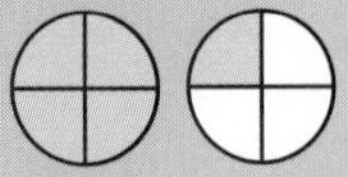

- If the top number is **smaller** than the bottom number, the fraction is **less than** 1 unit. $\frac{3}{4}$

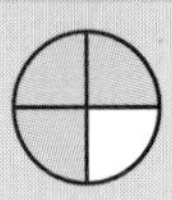

- If the top number is the **same** as the bottom number, the fraction **equals** 1 whole unit. The fraction $\frac{4}{4}$ equals 1 whole unit. $\frac{4}{4}$

- Remember, the bottom number tells how many parts each **unit** is divided into. The top number tells the number of shaded parts.

Part 2

Copy each fraction that is more than 1.

a. $\frac{6}{6}$ b. $\frac{3}{2}$ c. $\frac{56}{13}$ d. $\frac{80}{85}$ e. $\frac{1}{10}$ f. $\frac{7}{6}$

Part 3 Copy each fraction. Then write the letter of the picture for each fraction.

$\frac{7}{4}$ $\frac{3}{5}$ $\frac{1}{6}$ $\frac{6}{6}$

a. 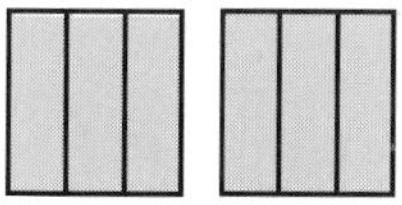

b. 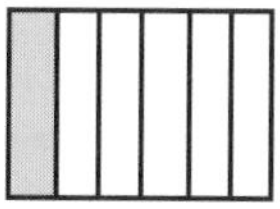

c. 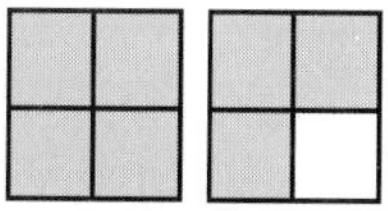

d. 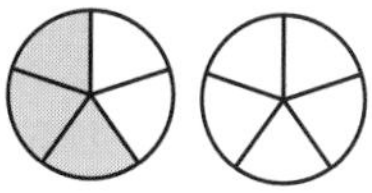

e. 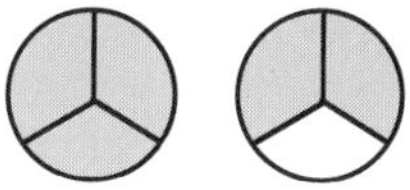

f. 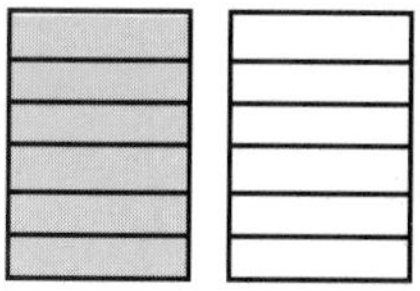

Part 4 Answer each question your teacher asks.

a. $\frac{9}{7} + \frac{3}{7} = \blacksquare$

b. $\frac{5}{10} - \frac{1}{10} = \blacksquare$

c. $\frac{20}{3} + \frac{56}{3} = \blacksquare$

Part 5 A student worked each problem. Copy and work the problems that have incorrect answers.

a. $\frac{13}{7} - \frac{2}{7} = \boxed{\frac{11}{7}}$

b. $\frac{3}{5} + \frac{24}{5} = \boxed{\frac{27}{10}}$

c. $\frac{1}{8} + \frac{7}{8} = \boxed{\frac{8}{8}}$

d. $\frac{14}{9} - \frac{8}{9} = \boxed{\frac{6}{0}}$

e. $\frac{12}{13} + \frac{6}{13} = \boxed{\frac{18}{26}}$

f. $\frac{5}{20} - \frac{4}{20} = \boxed{\frac{1}{20}}$

Part 6 Copy and work each problem.

a. $\frac{9}{7} + \frac{21}{7} = \blacksquare$

b. $\frac{14}{3} - \frac{14}{3} = \blacksquare$

c. $\frac{23}{12} - \frac{4}{12} = \blacksquare$

d. $\frac{7}{5} + \frac{5}{5} = \blacksquare$

e. $\frac{1}{32} - \frac{1}{32} = \blacksquare$

Part 7

- A lot of problems you'll work are based on number families. A number family is made up of three numbers that always go together in addition and subtraction problems.
- The family is made up of two small numbers and a big number. The big number is at the end of the arrow.

  small number — small number → big number

- Here's a family with the big number missing:

  40 — 160 → □

- To find the **big** number, you **add the small numbers.** 40 + 160 = 200. The big number is 200.

  40 — 160 → **200**

- Here's a family with a small number missing:

  □ — 160 → 200

- To find a missing **small** number, you **subtract.** You start with the big number and subtract the small number that is shown. The subtraction for this family is 200 – 160. So the missing small number is 40.

  40 — 160 → 200

- Here's a family with the other small number missing:

  40 — □ → 200

- You subtract to find that number. The subtraction for this family is 200 – 40. So the missing small number is 160.

  40 — **160** → 200

- Later you will use number families to figure out answers to very difficult word problems.

Part 8

Write the column problem and the answer for each number family.

a. 116 — 318 → ■

b. 207 — ■ → 470

c. ■ — 256 → 321

d. 867 — 135 → ■

Part 9

Figure out which of the problems the student worked incorrectly and work them correctly.

a. ■ 213 ⟶ 777

$$\begin{array}{r} \overset{1}{7}77 \\ +213 \\ \hline \boxed{990} \end{array}$$

b. 113 ■ ⟶ 141

$$\begin{array}{r} 1\overset{3}{\not4}{}^{1}1 \\ -113 \\ \hline \boxed{28} \end{array}$$

c. 176 408 ⟶ ■

$$\begin{array}{r} \overset{3}{\not4}{}^{1}08 \\ -176 \\ \hline \boxed{232} \end{array}$$

d. 263 ■ ⟶ 504

$$\begin{array}{r} 263 \\ -504 \\ \hline \boxed{361} \end{array}$$

e. 702 305 ⟶ ■

$$\begin{array}{r} 702 \\ +305 \\ \hline \boxed{1007} \end{array}$$

Part 10

Write the complete equation for each item.

a. 2 x ■ = 14 b. ■ x 5 = 20 c. ■ x 10 = 90

d. 6 x ■ = 30 e. 6 x ■ = 6

Independent Work

Part 11

Copy each problem and work it.

a. $\begin{array}{r} 45 \\ \times\ 61 \\ \hline \end{array}$ b. $\begin{array}{r} 127 \\ \times\ 9 \\ \hline \end{array}$ c. $\begin{array}{r} 560 \\ 103 \\ +\ 7026 \\ \hline \end{array}$ d. $\begin{array}{r} 33 \\ +\ 529 \\ \hline \end{array}$ e. $\begin{array}{r} 586 \\ -\ 399 \\ \hline \end{array}$

f. $6\overline{)744}$ g. $9\overline{)1647}$ h. $5\overline{)1170}$

Part J

b. $\frac{3}{5} + \frac{24}{5} = \boxed{\frac{27}{5}}$

d. $\frac{14}{9} - \frac{8}{9} = \boxed{\frac{6}{9}}$

e. $\frac{12}{13} + \frac{6}{13} = \boxed{\frac{18}{13}}$

Part K

a. $2 \times \boxed{7} = 14$ d. $6 \times \boxed{5} = 30$

b. $\boxed{4} \times 5 = 20$ e. $6 \times \boxed{1} = 6$

c. $\boxed{9} \times 10 = 90$

Lesson 2

Part 1 **Copy each fraction. Then write the letter of the picture for each fraction.**

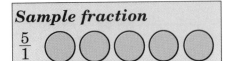

Sample fraction

$\frac{5}{1}$ 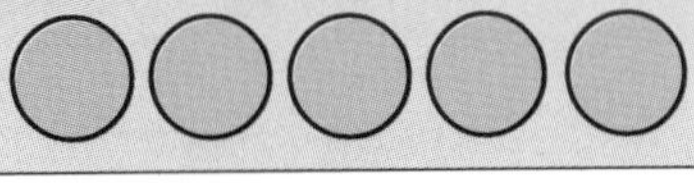

$\frac{3}{3}$ $\frac{6}{4}$ $\frac{3}{1}$ $\frac{3}{8}$ $\frac{1}{6}$

a. 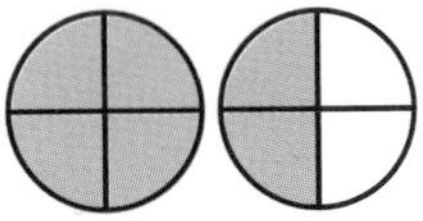

b. 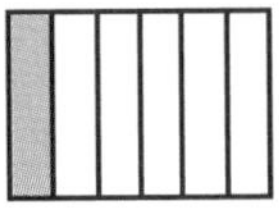

c. 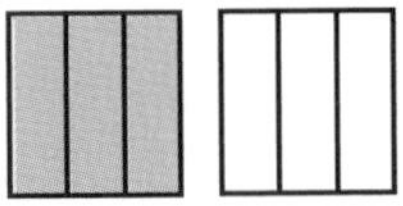

d. 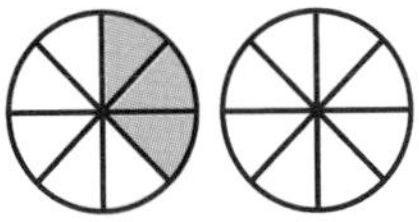

e. 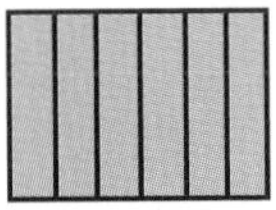

f. 

Part 2 **Copy and work each problem.**

a. $\frac{15}{8} - \frac{3}{8} =$ ■

b. $\frac{2}{5} + \frac{9}{5} =$ ■

c. $\frac{3}{8} - \frac{3}{8} =$ ■

d. $\frac{27}{7} - \frac{6}{7} =$ ■

e. $\frac{6}{12} - \frac{1}{12} =$ ■

f. $\frac{5}{9} + \frac{9}{9} =$ ■

Part 3 **Write the column problem and the answer for each family.**

a. 248 ■ → 371

b. 567 98 → ■

c. ■ 104 → 286

d. 83 ■ → 179

Part 4 **Write the complete equation for each item.**

- Any number x 1 = the number.
- Any number x 0 = 0.

a. 15 x ■ = 15

b. 7 x ■ = 0

c. 9 x ■ = 90

d. ■ x 9 = 0

e. ■ x 5 = 35

f. ■ x 5 = 5

g. ■ x 8 = 48

Part 5 **A student worked each problem. Copy and work the problems that have incorrect answers.**

a. $\frac{12}{12} - \frac{5}{12} = \boxed{\frac{7}{0}}$

b. $\frac{26}{1} + \frac{5}{1} = \boxed{\frac{31}{1}}$

c. $\frac{17}{5} - \frac{17}{5} = \boxed{\frac{0}{5}}$

d. $\frac{4}{9} + \frac{4}{9} = \boxed{\frac{8}{9}}$

e. $\frac{19}{6} - \frac{6}{6} = \boxed{\frac{13}{6}}$

f. $\frac{4}{14} + \frac{18}{14} = \boxed{\frac{22}{28}}$

Part 6

- Here's a word problem that you can solve by multiplication:

  If each of the bricks weighs 64 ounces, how much do 12 bricks weigh?

- The first part of the problem tells you that 1 brick weighs 64 ounces.
- The problem asks about the weight of more than 1 brick.
- To find the weight of more than 1 brick, you multiply.
- You find the weight of 12 bricks by working the problem:

$$\begin{array}{r} \mathbf{64} \\ \underline{\mathbf{x\,12}} \end{array}$$

- Here's a problem you can solve using division:

  If 15 identical tiles weigh 390 grams, how much does each tile weigh?

- The problem tells what 15 tiles weigh and asks about the weight of 1 tile.
- To find the weight of 1 tile, you divide. $\mathbf{15\overline{)390}}$
- Remember, if the problem tells about **1** and asks about **more** than 1, you work a **multiplication** problem. If the problem tells about **more** than 1 and asks about **1,** you work a **division** problem.

Part 7

Write multiplication or division for each problem. Then work each problem.

a. Each stamp costs 19 cents. How much do 9 stamps cost?

b. There were 678 toothpicks in groups that were the same size. There were 6 groups. How many toothpicks were in each group?

c. 16 worms were arranged in equal-sized groups. There were 8 groups of worms. How many worms were there in each group?

d. A machine made 7 buttonholes each minute. How many buttonholes would the machine make in 45 minutes?

Part 8

Copy each fraction. After each fraction, write more than 1, less than 1 or equals 1.

a. $\frac{12}{11}$ b. $\frac{2}{2}$ c. $\frac{3}{2}$ d. $\frac{30}{30}$ e. $\frac{8}{1}$ f. $\frac{1}{8}$

Independent Work

Part 9

Copy each problem and work it. Box the answer.

a. $\begin{array}{r} 387 \\ \underline{\times\ 20} \end{array}$ b. $\begin{array}{r} 463 \\ \underline{\times\ 16} \end{array}$ c. $7\overline{)728}$ d. $6\overline{)414}$ e. $\begin{array}{r} 51 \\ 306 \\ \underline{+\ 915} \end{array}$ f. $\begin{array}{r} 876 \\ \underline{-\ 211} \end{array}$ g. $\begin{array}{r} 401 \\ \underline{-\ 386} \end{array}$ h. $\begin{array}{r} 5 \\ 19 \\ \underline{+\ 234} \end{array}$

Part J

a. $\frac{15}{8} - \frac{3}{8} = \boxed{\frac{12}{8}}$

b. $\frac{2}{5} + \frac{9}{5} = \boxed{\frac{11}{5}}$

c. $\frac{3}{8} - \frac{3}{8} = \boxed{\frac{0}{8}}$

d. $\frac{27}{7} - \frac{6}{7} = \boxed{\frac{21}{7}}$

e. $\frac{6}{12} - \frac{1}{12} = \boxed{\frac{5}{12}}$

f. $\frac{5}{9} + \frac{9}{9} = \boxed{\frac{14}{9}}$

Part K

a. $15 \times \boxed{1} = 15$

b. $7 \times \boxed{0} = 0$

c. $9 \times \boxed{10} = 90$

d. $\boxed{0} \times 9 = 0$

e. $\boxed{7} \times 5 = 35$

f. $\boxed{1} \times 5 = 5$

g. $\boxed{6} \times 8 = 48$

Lesson 3

Part 1 **Figure out the missing number for each family.**

a. 110 89 ⟶ ■

b. 529 ■ ⟶ 680

c. ■ 350 ⟶ 458

d. 65 128 ⟶ ■

Part 2 **Figure out which items the student worked incorrectly and work them correctly.**

a. ■ 115 ⟶ 238

$$\begin{array}{r} 238 \\ +\,115 \\ \hline \boxed{353} \end{array}$$

b. 601 298 ⟶ ■

$$\begin{array}{r} 601 \\ -\,298 \\ \hline \boxed{303} \end{array}$$

c. 482 ■ ⟶ 725

$$\begin{array}{r} 725 \\ -\,482 \\ \hline \boxed{243} \end{array}$$

d. 84 426 ⟶ ■

$$\begin{array}{r} 84 \\ +\,426 \\ \hline \boxed{510} \end{array}$$

e. ■ 523 ⟶ 891

$$\begin{array}{r} 523 \\ -\,891 \\ \hline \boxed{372} \end{array}$$

f. 350 ■ ⟶ 1075

$$\begin{array}{r} 1075 \\ +\,350 \\ \hline \boxed{1425} \end{array}$$

Part 3 **Copy each fraction. After each fraction, write more than 1, less than 1 or equals 1.**

a. $\frac{13}{14}$ b. $\frac{2}{3}$ c. $\frac{24}{24}$ d. $\frac{20}{4}$ e. $\frac{4}{4}$ f. $\frac{2}{9}$

Part 4

Write multiplication or division for each problem. Then work each problem.

a. Roy wants to make stacks of magazines that have the same number in each stack. He wants to make 5 stacks. He has 265 magazines. How many magazines will be in each stack?

b. Edna has 14 identical coins. Each coin is worth $50. How much are all the coins worth?

c. Baseball cards weigh 3 grams each. Tim has 89 cards. What is the weight of all of his cards?

d. A cake is divided into 8 equal pieces. If the entire cake weighs 96 ounces, how much does each piece weigh?

Part 5

Copy all the problems you can work the way they are written and work them.

a. $\frac{2}{3} + \frac{2}{5} = \blacksquare$

b. $\frac{17}{4} - \frac{5}{4} = \blacksquare$

c. $\frac{4}{8} + \frac{8}{8} = \blacksquare$

d. $\frac{6}{7} - \frac{7}{6} = \blacksquare$

e. $\frac{9}{3} + \frac{14}{3} = \blacksquare$

f. $\frac{8}{4} + \frac{2}{6} = \blacksquare$

g. $\frac{5}{8} - \frac{5}{7} = \blacksquare$

h. $\frac{27}{2} - \frac{27}{2} = \blacksquare$

Part 6

Write the complete equation for each item.

Any number + 0 = the number.

a. $9 + \blacksquare = 15$

b. $6 + \blacksquare = 16$

c. $6 \times \blacksquare = 0$

d. $\blacksquare \times 19 = 19$

e. $\blacksquare + 19 = 19$

f. $\blacksquare \times 19 = 0$

g. $17 + \blacksquare = 17$

h. $5 \times \blacksquare = 5$

Independent Work

Part 7 **Work each problem. Write problems a through f as column problems.**

a. 312 – 181 = [redacted]

b. 267 + 48 + 183 = [redacted]

c. 411 – 298 = [redacted]

d. 14 x 12 = [redacted]

e. 381 x 17 = [redacted]

f. 406 x 40 = [redacted]

g. $6\overline{)630}$

h. $2\overline{)890}$

i. $4\overline{)2112}$

Part 8 **Copy each fraction. For each fraction, write the letter of the picture that shows the fraction.**

$\frac{3}{4}$ $\frac{4}{4}$ $\frac{3}{2}$ $\frac{1}{2}$

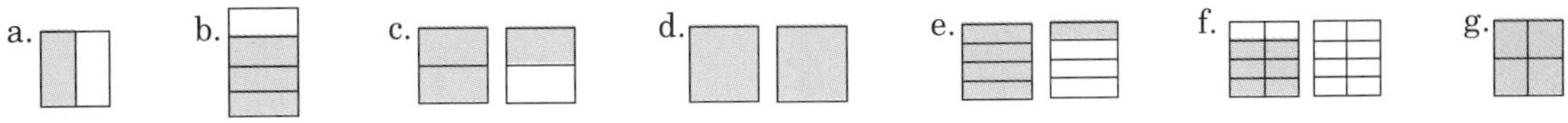

Lesson 4

Part 1 **Copy all the problems that can be worked the way they are written and work them.**

a. $\frac{35}{4} - \frac{3}{4} = \blacksquare$ b. $\frac{8}{5} + \frac{8}{3} = \blacksquare$ c. $\frac{12}{5} - \frac{0}{5} = \blacksquare$

d. $\frac{2}{7} - \frac{1}{7} = \blacksquare$ e. $\frac{3}{8} + \frac{5}{8} = \blacksquare$ f. $\frac{7}{3} + \frac{9}{1} = \blacksquare$

Part 2

- When you add or subtract fractions, you don't add or subtract the bottom numbers. You just copy the bottom number in the answer.
- When you **multiply** fractions, you do not follow the same procedure.
  - ✔ You multiply the top numbers together and write the answer on top.
  - ✔ You multiply the bottom numbers together and write the answer on the bottom.
- Here's: $\frac{2}{5} \times \frac{3}{4} = \blacksquare$
- On top, you work the problem 2 x 3. You write the answer on top. $\frac{2}{5} \times \frac{3}{4} = \frac{6}{\blacksquare}$
- On the bottom, you work the problem 5 x 4 and write the answer on the bottom. $\frac{2}{5} \times \frac{3}{4} = \frac{6}{20}$

Part 3 **Say each problem your teacher names.**

a. $\frac{2}{3} \times \frac{7}{8} = \blacksquare$ b. $\frac{1}{5} \times \frac{9}{3} = \blacksquare$ c. $\frac{5}{4} \times \frac{2}{6} = \blacksquare$

Part 4

A student worked each problem. Copy and work the problems that have incorrect answers.

a. $\frac{3}{4} \times \frac{1}{5} = \boxed{\frac{3}{9}}$ b. $\frac{4}{9} \times \frac{2}{1} = \boxed{\frac{8}{9}}$ c. $\frac{4}{5} \times \frac{8}{5} = \boxed{\frac{32}{5}}$

d. $\frac{1}{7} \times \frac{8}{1} = \boxed{\frac{9}{8}}$ e. $\frac{10}{3} \times \frac{5}{7} = \boxed{\frac{50}{21}}$ f. $\frac{6}{9} \times \frac{1}{9} = \boxed{\frac{6}{81}}$

Part 5

Copy and work each item.

a. $\frac{7}{3} \times \frac{4}{3} =$ ■ b. $\frac{8}{11} \times \frac{2}{2} =$ ■

c. $\frac{3}{6} \times \frac{4}{2} =$ ■ d. $\frac{5}{8} \times \frac{9}{2} =$ ■

Part 6

Copy the families that are correct and figure out the missing number for those families. Don't work problems for the families that are impossible.

a. 58 ■ → 85 b. 76 23 → ■ c. ■ 200 → 168

d. 64 128 → ■ e. 73 ■ → 49 f. 48 ■ → 56

Part 7

- Some problems show multiplication with a times sign. Here's a problem written with a times sign: **3 x 6 = ■**
- Sometimes, parentheses are used to show multiplication. Here's the same problem written with parentheses: **3 (6) = ■**
  You read that problem the same way you read the other problem: 3 times 6.
- You can also use parentheses to show fractions that are multiplied. Here's $\frac{3}{4}$ times $\frac{5}{8}$: $\frac{3}{4}\left(\frac{5}{8}\right) =$ ■

Part 8 Write the complete equation for each item.

Sample problem 1	*Sample problem 2*
$4(■) = 20$	$9(8) = ■$

a. $6(■) = 48$

b. $4(8) = ■$

c. $20(■) = 20$

d. $15(3) = ■$

e. $5(■) = 50$

Part 9 Copy all the fractions that are more than 1 or that equal 1. Then circle the fractions that equal 1.

$\frac{7}{7}$ $\frac{3}{1}$ $\frac{1}{3}$ $\frac{2}{3}$ $\frac{3}{2}$ $\frac{3}{3}$ $\frac{1}{5}$ $\frac{7}{3}$ $\frac{3}{7}$

Part 10 Work each item. Remember the unit name.

a. A typist typed 384 words in 6 minutes. How many words did the typist type each minute?

b. Another typist typed at the rate of 75 words each minute. That typist typed for 11 minutes. How many words did the typist type?

c. 56 bottles are put into 7 cartons. Each carton holds the same number of bottles. How many bottles are in each carton?

d. Each time Mrs. Smith goes on a bike ride, she travels 28 miles. During April, she went on 4 bike rides. How far did she go in all?

Independent Work

Part 11 Copy and work each item.

a. $2\overline{)5030}$

b. $4\overline{)5032}$

c. $5\overline{)5030}$

d. $5\overline{)5060}$

e. $\begin{array}{r} 380 \\ \times \quad 4 \\ \hline \end{array}$

f. $\begin{array}{r} 21 \\ \times 15 \\ \hline \end{array}$

Part 12 Copy each item and work it.

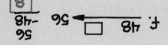

a. $\frac{15}{7} - \frac{12}{7} =$ ■

b. $\frac{3}{8} + \frac{11}{8} =$ ■

c. $\frac{5}{5} - \frac{5}{5} =$ ■

d. $\frac{13}{17} - \frac{8}{17} =$ ■

e. $\frac{1}{2} + \frac{7}{2} =$ ■

Part 13 Copy and work each item.

a. ■ + 9 = 15

b. ■ x 4 = 8

c. ■ x 6 = 6

d. ■ x 2 = 8

e. ■ + 4 = 8

f. 5 x ■ = 15

g. 12 – ■ = 5

h. 7 x ■ = 0

Part J

a. $\frac{7}{3} \times \frac{4}{3} = \frac{28}{9}$

b. $\frac{8}{11} \times \frac{2}{2} = \frac{16}{22}$

c. $\frac{3}{6} \times \frac{4}{2} = \frac{12}{12}$

d. $\frac{5}{8} \times \frac{9}{2} = \frac{45}{16}$

Part K

a. 58 □ → 85; 85 − 58 = 27

b. 76 23 → □; 76 + 23 = 99

d. 64 128 → □; 64 + 128 = 192

f. 48 □ → 56; 56 − 48 = 8

Part L

a. $6\overline{)384}$ 64 words

b. 75 × 11: 75 + 750 = 825 words

c. $7\overline{)56}$ 8 bottles

d. 28 × 4 = 112 miles

Lesson 5

Part 1

A student worked each problem. Copy and work the problems that have incorrect answers.

a. $\frac{7}{3}\left(\frac{7}{3}\right) = \boxed{\frac{49}{9}}$ b. $\frac{5}{3}\left(\frac{0}{3}\right) = \boxed{\frac{5}{9}}$ c. $\frac{4}{5}\left(\frac{10}{5}\right) = \boxed{\frac{40}{5}}$ d. $\frac{8}{10}\left(\frac{2}{2}\right) = \boxed{\frac{16}{20}}$

Part 2

Copy and work each item.

a. $\frac{7}{10}\left(\frac{3}{8}\right) = ■$ b. $\frac{1}{2}\left(\frac{1}{5}\right) = ■$ c. $\frac{10}{7}\left(\frac{2}{7}\right) = ■$ d. $\frac{8}{13}\left(\frac{8}{1}\right) = ■$

Part 3

Copy the families that are correct and figure out the missing number for those families. Don't work problems for the families that are impossible.

a. 64 87 → ■ b. ■ 102 → 95 c. ■ 487 → 298

d. ■ 45 → 386 e. 684 27 → ■ f. 267 ■ → 159

Part 4

Copy all the fractions that are more than 1 or that equal 1. Then circle the fractions that equal 1.

$\frac{12}{13}$ $\frac{6}{5}$ $\frac{5}{5}$ $\frac{19}{31}$ $\frac{9}{13}$ $\frac{80}{72}$ $\frac{8}{17}$ $\frac{2}{1}$ $\frac{20}{20}$ $\frac{20}{100}$

Part 5

For each item, make a number family. Then figure out the missing value.

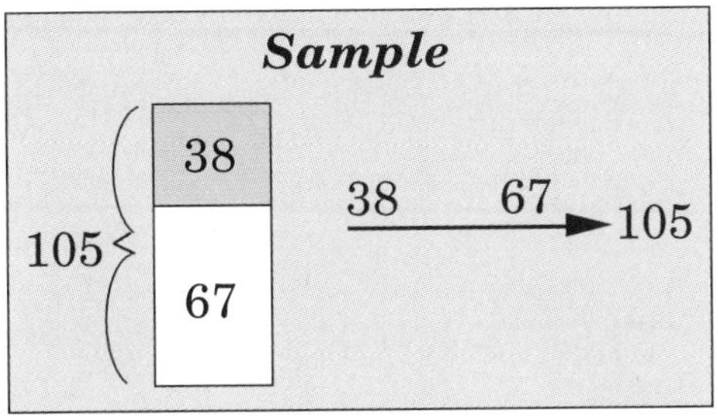

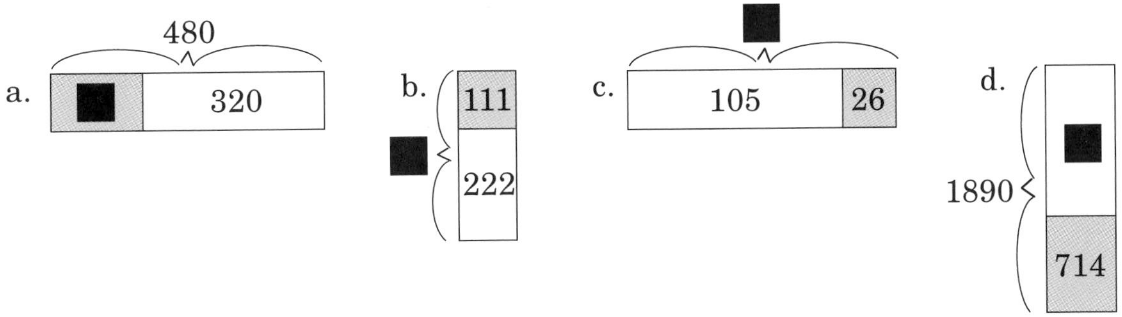

Part 6

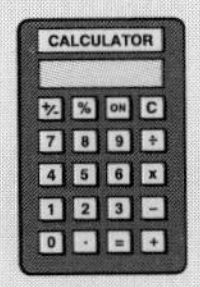

You can work multiplication problems that have the middle number missing.

- Here's: **4 (   ) = 28**
- You use the multiplication fact that has the numbers 4 and 28: **4 (7) = 28**

You can also work these problems as division problems.

- Here's the same problem written as a division problem: $4\overline{)28}$ with 7 written above

- When problems like these have large numbers, rewrite them as division problems.
  - Here's: **45 ([redacted]) = 3690**
  - That's: [redacted] above **$45\overline{)3690}$**
- When you work that problem, you'll know what you'd multiply 45 by to get 3690.

Part 7

Copy each problem. Work the division problems and write the answers.

a. $49(\blacksquare) = 343$ b. $46(\blacksquare) = 2392$

c. $56(\blacksquare) = 1904$ d. $28(\blacksquare) = 532$

Part 8

Work each item. Remember the unit name.

a. Each chicken on the Foster Farm laid 11 eggs. There were 58 chickens on the farm. How many eggs did the chickens lay in all?

b. A pie was divided into 9 equal-sized parts. The entire pie weighed 45 ounces. How many ounces did each piece weigh?

c. A truck made 4 stops each day. The truck made stops on 16 days. How many stops did the truck make in all?

d. 64 books were put on 4 shelves. The same number of books were on each shelf. How many books were on each shelf?

Independent Work

Part 9

Work all the problems that can be worked the way they are written.

a. $\frac{10}{15} - \frac{10}{10} = \blacksquare$ d. $\frac{4}{7} - \frac{4}{0} = \blacksquare$

b. $\frac{4}{7} + \frac{20}{7} = \blacksquare$ e. $\frac{13}{9} + \frac{13}{9} = \blacksquare$

c. $\frac{15}{16} - \frac{1}{16} = \blacksquare$ f. $\frac{19}{5} - \frac{18}{5} = \blacksquare$

Part 10

Copy and work each item.

a. $\begin{array}{r} 64 \\ \times\ 75 \\ \hline \end{array}$ b. $\begin{array}{r} 37 \\ +\ 239 \\ \hline \end{array}$ c. $8\overline{)1040}$ d. $3\overline{)6903}$

e. $\begin{array}{r} 900 \\ -\ 684 \\ \hline \end{array}$ f. $\begin{array}{r} 870 \\ \times\ 28 \\ \hline \end{array}$ g. $\begin{array}{r} 461 \\ 3985 \\ +\ 79 \\ \hline \end{array}$

Part 11

Copy each family and figure out the missing number.

a. $59 \quad \blacksquare \longrightarrow 320$

b. $256 \quad 81 \longrightarrow \blacksquare$

c. $\blacksquare \quad 148 \longrightarrow 176$

Part 12

For each item, write the missing value.

a. $5 \times \blacksquare = 30$ d. $15 - \blacksquare = 1$ f. $\blacksquare \times 9 = 54$

b. $5 + \blacksquare = 30$ e. $7 \times \blacksquare = 70$ g. $\blacksquare \times 3 = 24$

c. $15 - \blacksquare = 0$

Part J

a. $\frac{7}{10}\left(\frac{3}{8}\right) = \boxed{\frac{21}{80}}$ b. $\frac{1}{2}\left(\frac{1}{5}\right) = \boxed{\frac{1}{10}}$ c. $\frac{10}{7}\left(\frac{2}{7}\right) = \boxed{\frac{20}{49}}$ d. $\frac{8}{13}\left(\frac{8}{1}\right) = \boxed{\frac{64}{13}}$

Part K

a. 64 87 → □ $64 + 87 = \boxed{151}$

d. □ 45 → 386 $386 - 45 = \boxed{341}$

e. 684 27 → □ $684 + 27 = \boxed{711}$

Part L

a. 58×11: 58 + 580 = **638 eggs**

b. $9\overline{)45}$ **5 ounces**

c. 16×4 = **64 stops**

d. $4\overline{)64}$ **16 books**

Lesson 6

Part 1 Copy and rework each item that is incorrect.

a. $\frac{1}{2} \times \frac{1}{2} = \boxed{\frac{2}{2}}$

b. $\frac{12}{20} - \frac{3}{20} = \boxed{\frac{9}{20}}$

c. $\frac{2}{3} \times \frac{3}{2} = \boxed{\frac{6}{1}}$

d. $\frac{15}{3} - \frac{8}{3} = \boxed{\frac{7}{0}}$

e. $\frac{8}{25} + \frac{20}{25} = \boxed{\frac{28}{25}}$

f. $\frac{3}{5} + \frac{7}{5} = \boxed{\frac{21}{25}}$

g. $\frac{0}{9}\left(\frac{3}{9}\right) = \boxed{\frac{0}{81}}$

Part 2 Copy and work each item.

a. $\frac{5}{8} + \frac{5}{8} = ■$

b. $\frac{5}{8} \times \frac{5}{8} = ■$

c. $\frac{5}{8} - \frac{5}{8} = ■$

d. $\frac{17}{3} \times \frac{1}{5} = ■$

e. $\frac{3}{7} \times \frac{6}{7} = ■$

f. $\frac{6}{4} - \frac{2}{4} = ■$

Part 3 Copy and complete rows a through d of the table.

	Multiplication	Division
Sample	**23 (41) = 943**	**$23\overline{)943}$ = 41**
a.	36 (■) = 684	
b.	14 (■) = 406	
c.	37 (■) = 888	
d.	45 (■) = 1575	

Part 4

Copy each fraction, write an equal sign and the whole number it equals.

Sample $\frac{15}{5} = \boxed{3}$

a. $\frac{12}{6}$　　b. $\frac{18}{3}$　　c. $\frac{43}{1}$　　d. $\frac{24}{8}$

Part 5

For each item, make a number family. Then figure out the missing value.

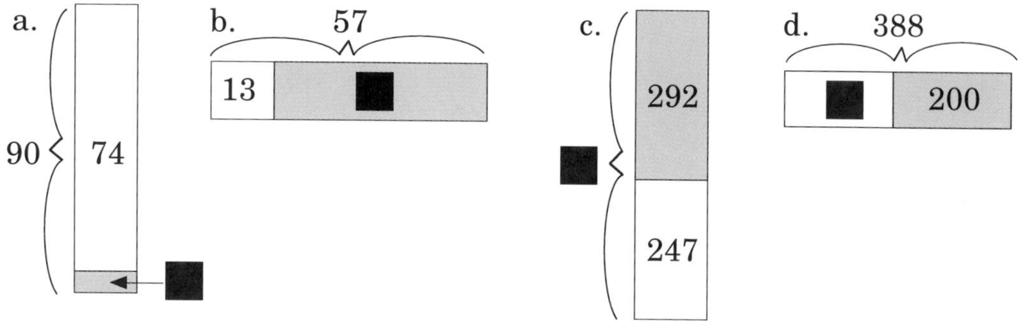

Part 6

- You've worked problems like this one: $5(\blacksquare) = 30$

  You multiply 5 by 6. $5(6) = 30$
- You can work similar problems that involve fractions.
  - Here's: $\frac{5}{3}\left(\frac{\blacksquare}{\blacksquare}\right) = \frac{30}{12}$
- You can write the missing value as a fraction. You do that by working the problem on the top and the problem on the bottom.
- Here's the equation that shows the fraction you multiply $\frac{5}{3}$ by to get $\frac{30}{12}$. $\frac{5}{3}\left(\frac{6}{4}\right) = \frac{30}{12}$

Part 7 Copy each problem and write the missing fraction.

a. $\frac{2}{7}\left(\frac{\blacksquare}{\blacksquare}\right) = \frac{6}{14}$ b. $\frac{7}{4}\left(\frac{\blacksquare}{\blacksquare}\right) = \frac{35}{32}$ c. $\frac{1}{2}\left(\frac{\blacksquare}{\blacksquare}\right) = \frac{2}{2}$

d. $\frac{5}{9}\left(\frac{\blacksquare}{\blacksquare}\right) = \frac{15}{27}$ e. $\frac{10}{5}\left(\frac{\blacksquare}{\blacksquare}\right) = \frac{30}{45}$

Part 8 Copy each family that you can work and figure out the missing number.

a. $\xrightarrow{264 \quad 19} \blacksquare$ b. $\xrightarrow{\blacksquare \quad 183} 147$ c. $\xrightarrow{436 \quad \blacksquare} 390$

d. $\xrightarrow{25 \quad 98} \blacksquare$ e. $\xrightarrow{613 \quad \blacksquare} 479$ f. $\xrightarrow{213 \quad \blacksquare} 286$

Independent Work

Part 9 Copy each fraction that is equal to 1 or more than 1. Circle fractions that equal 1.

$\frac{26}{27}$ $\frac{8}{3}$ $\frac{5}{4}$ $\frac{32}{30}$ $\frac{12}{12}$ $\frac{3}{8}$ $\frac{8}{8}$

Part 10 Write the complete equation for each item.

a. $4 \times \blacksquare = 24$ b. $\blacksquare \times 3 = 27$ c. $17 - \blacksquare = 17$ d. $\blacksquare + 10 = 20$ e. $15 \times 2 = \blacksquare$

f. $\blacksquare \times 64 = 64$ g. $\blacksquare + 7 = 15$ h. $\blacksquare \times 7 = 21$ i. $36 \times \blacksquare = 0$

Part 11 Work all the problems that can be worked the way they are written.

a. $\frac{14}{14} - \frac{1}{14} = \blacksquare$ d. $\frac{90}{5} - \frac{90}{5} = \blacksquare$

b. $\frac{3}{8} + \frac{15}{8} = \blacksquare$ e. $\frac{12}{20} + \frac{8}{20} = \blacksquare$

c. $\frac{8}{5} - \frac{8}{8} = \blacksquare$

Part 12 Copy each problem and work it.

a. $\frac{3}{7}\left(\frac{9}{7}\right) = \blacksquare$ b. $\frac{3}{1}\left(\frac{5}{1}\right) = \blacksquare$

c. $\frac{9}{8} \times \frac{0}{8} = \blacksquare$ d. $\frac{12}{1} \times \frac{1}{12} = \blacksquare$

Part 13

Work each problem. Remember the unit name.

a. John had 640 marbles. He put an equal number of marbles in each of 8 bags. How many marbles were in each bag?

b. Each spider had 8 legs. There were 41 spiders. How many legs were there?

c. 400 people went in cars. The same number of people were in each car. There were 80 cars. How many people were in each car?

d. There were 64 cases of beans in a warehouse. Each case contained 15 cans. How many cans of beans were in the warehouse?

Part J

a. $\frac{5}{8} + \frac{5}{8} = \boxed{\frac{10}{8}}$

b. $\frac{5}{8} \times \frac{5}{8} = \boxed{\frac{25}{64}}$

c. $\frac{5}{8} - \frac{5}{8} = \boxed{\frac{0}{8}}$

d. $\frac{17}{3} \times \frac{1}{5} = \boxed{\frac{17}{15}}$

e. $\frac{3}{7} \times \frac{6}{7} = \boxed{\frac{18}{49}}$

f. $\frac{6}{4} - \frac{2}{4} = \boxed{\frac{4}{4}}$

Part K

a. 264 19 → $264 + 19 = \boxed{283}$

d. 25 98 → $25 + 98 = \boxed{123}$

f. 213 □ 286 → $286 - 213 = \boxed{73}$

I could figure out the number of legs if I could ever get these spiders to stay still.
Dr. Thompson, there's an easier way.
41
x 8

Lesson 7

Part 1 **Copy each item and write the missing fraction.**

a. $\frac{3}{5}(■) = \frac{18}{30}$ b. $\frac{5}{1}(■) = \frac{10}{1}$ c. $\frac{6}{5}(■) = \frac{24}{20}$

d. $\frac{4}{7}(■) = \frac{4}{14}$ e. $\frac{2}{5}(■) = \frac{6}{10}$

Part 2 **For each item, make a number family with three names and two numbers. Figure out the missing number. Write the answer as a number and unit name. Box the answer.**

Sample problem 1

A bar has a shaded part and an unshaded part. The whole bar is 381 inches long. The shaded part is 110 inches long. How long is the unshaded part?

shaded	**unshaded**	**bar**
110	■	→ 381

$$\begin{array}{r} 381 \\ -\ 110 \\ \hline \boxed{271 \text{ inches}} \end{array}$$

Sample problem 2

A bar has an unshaded part and a shaded part. The unshaded part is 56 feet long. The shaded part is 33 feet long. How long is the whole bar?

unshaded	**shaded**	**bar**
56	33	→ ■

$$\begin{array}{r} 56 \\ +\ 33 \\ \hline \boxed{89 \text{ feet}} \end{array}$$

a. A bar has a shaded part and an unshaded part. The unshaded part is 49 meters long. The whole bar is 400 meters long. How long is the shaded part?

b. The unshaded part of a bar is 306 centimeters long. The shaded part of the bar is 132 centimeters. How long is the whole bar?

c. A whole bar is 398 centimeters long. The shaded part of the bar is 311 centimeters. How long is the unshaded part of the bar?

d. The whole bar is 216 meters long. The unshaded part is 133 meters long. How long is the shaded part?

e. The shaded part of a bar weighs 98 pounds. The unshaded part weighs 51 pounds. How much does the whole bar weigh?

Part 3

Write the equation to show the whole number each fraction equals.

a. $\frac{7}{1}$ b. $\frac{14}{2}$ c. $\frac{63}{9}$

d. $\frac{9}{9}$ e. $\frac{36}{9}$ f. $\frac{70}{10}$

Part 4

Copy the table and complete it.

	Multiplication	Division
a.	$6(\blacksquare) = 564$	$\overline{)\quad}$
b.	$72(\blacksquare) = 792$	
c.	$3(\blacksquare) = 918$	
d.	$16(\blacksquare) = 704$	

Part 5

Copy each item and work it.

Sample problem $2 \times 4 \times 3 = \blacksquare$

a. $\frac{4}{7} + \frac{9}{7} - \frac{5}{7} = \blacksquare$ b. $\frac{2}{3} \times \frac{2}{2} \times \frac{9}{1} = \blacksquare$ c. $\frac{12}{8} - \frac{4}{8} + \frac{10}{8} = \blacksquare$

d. $\frac{5}{4}\left(\frac{2}{4}\right)\left(\frac{1}{2}\right) = \blacksquare$ e. $\frac{14}{6} + \frac{7}{6} + \frac{7}{6} = \blacksquare$ f. $\frac{1}{2} \times \frac{1}{3} \times \frac{1}{4} = \blacksquare$

Independent Work

Part 6 **Copy the fractions that are less than 1 or equal to 1. Circle the fractions that are equal to 1.**

$\frac{14}{40}$ $\frac{40}{39}$ $\frac{15}{16}$ $\frac{39}{39}$ $\frac{1}{1}$ $\frac{5}{8}$ $\frac{1}{2}$ $\frac{3}{7}$ $\frac{7}{7}$ $\frac{3}{1}$

Part 7 **Copy each family. Figure out the missing number.**

a. ■ 189 → 264

b. 206 13 → ■

c. 111 ■ → 215

Part 8 **Write the complete equation for each item.**

a. ■ x 8 = 40

b. 6 x 1 = ■

c. 5 x ■ = 60

d. ■ + 1 = 40

e. ■ – 1 = 40

f. ■ x 1 = 40

g. 8 x 0 = ■

h. 6 x ■ = 36

i. 6 x 8 = ■

j. ■ x 40 = 0

Part 9 **Copy and work each problem.**

a. $4\overline{)356}$

b. $2\overline{)356}$

c. $5\overline{)705}$

d. $9\overline{)297}$

e. 56 x 38

f. 46 x 30

g. 19 x 81

Part 10 **Work each problem. Remember the unit name.**

a. There are 12 bottles in each carton. How many bottles are there in 9 cartons?

b. A bus travels 130 miles each day. How far does the bus travel in 3 days?

c. A company divides $420 among 10 employees so that each employee receives the same amount of money. How much does each employee receive?

d. 148 shells were placed in 4 groups that are the same size. How many shells were in each group?

Lesson 8

Part 1 **Copy and work each problem.**

a. $\frac{4}{2}\left(\frac{7}{6}\right) = \blacksquare$ b. $\frac{3}{1}\left(\blacksquare\right) = \frac{15}{1}$ c. $\frac{9}{4}\left(\blacksquare\right) = \frac{81}{12}$ d. $\frac{3}{8}\left(\frac{6}{8}\right) = \blacksquare$

Part 2 **For each item, make a number family with three names and two numbers.**

a. A bar has a shaded part and an unshaded part. The shaded part weighs 66 pounds. The unshaded part weighs 123 pounds. How much does the whole bar weigh?

b. The unshaded part of a bar is 140 inches long. The entire bar is 200 inches long. How long is the part that is shaded?

c. A bar weighs 89 tons. The shaded part weighs 14 tons. How much does the unshaded part weigh?

d. The unshaded part of a bar is 41 centimeters long. The entire bar is 203 centimeters long. How long is the shaded part of the bar?

e. The unshaded part of a bar is 600 meters long. The shaded part is 513 meters long. How long is the whole bar?

I can work this problem the way it's written–I just can't get the right answer.

Part 3

- Some fractions are equivalent.
- If fractions are equivalent, you can write an equal sign between them.

- $\frac{1}{2}$ and $\frac{4}{8}$ are equivalent.
- If fractions are equivalent, pictures of the fractions show exactly the same area that is shaded.

$$\frac{1}{2} = \frac{4}{8}$$

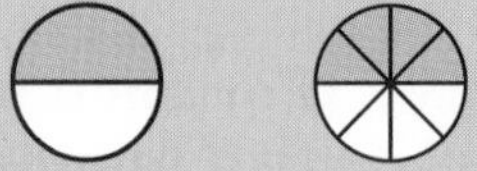

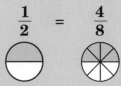

- You can figure out whether any two fractions are equivalent by finding the fraction you multiply the first fraction by to get the other fraction.
- If you multiply by a fraction that equals 1, the fractions you started with are equivalent.

- Here's a problem: $\frac{1}{2}(\blacksquare) = \frac{4}{8}$

$$\frac{1}{2}\left(\frac{4}{4}\right) = \frac{4}{8}$$

So: $\frac{1}{2} = \frac{4}{8}$

Part 4

For each item, write the complete equation. If the fractions are equivalent, write a simple equation below.

a. $\frac{2}{5}, \frac{8}{10}$

b. $\frac{4}{5}, \frac{8}{10}$

c. $\frac{3}{4}, \frac{15}{20}$

d. $\frac{4}{1}, \frac{12}{3}$

e. $\frac{6}{5}, \frac{36}{35}$

Part 5

Copy the table and complete it.

	Multiplication	Division
a.	56 x 12 = $\blacksquare$	12
b.	13 x 61 = $\blacksquare$	61
c.	48 x $\blacksquare$ = 1008	
d.	74 x $\blacksquare$ = 962	

Part 6 **For each item, write the fraction equation.**

Sample problem	$1 + 1 + 1 = 3$
	$\frac{5}{5} + \frac{5}{5} + \frac{5}{5} = \frac{15}{5}$

a. $1 + 1 = 2$

$\frac{■}{7} + \frac{■}{7} = ■$

b. $1 + 1 + 1 + 1 = 4$

$\frac{■}{3} + \frac{■}{3} + \frac{■}{3} + \frac{■}{3} = ■$

c. $1 + 1 + 1 = 3$

$\frac{■}{10} + \frac{■}{10} + \frac{■}{10} = ■$

d. $1 + 1 + 1 = 3$

$\frac{■}{8} + \frac{■}{8} + \frac{■}{8} = ■$

Independent Work

Part 7 **Copy the fractions that are more than 1 or equal to 1. Circle the fractions that are more than 1.**

$\frac{3}{3}$ $\frac{12}{15}$ $\frac{11}{11}$ $\frac{9}{8}$ $\frac{5}{6}$ $\frac{112}{200}$ $\frac{56}{80}$ $\frac{1}{7}$ $\frac{7}{7}$ $\frac{7}{1}$

Part 8 **Copy the families that you can work and work them.**

a. 56 ■ → 94

b. ■ 240 → 186

c. 56 94 → ■

d. ■ 94 → 156

e. 67 ■ → 48

Part 9 **Work all the problems that can be worked the way they are written.**

a. $\frac{3}{7} \times \frac{1}{4} =$ ■

b. $\frac{5}{5} + \frac{5}{8} =$ ■

c. $\frac{1}{9} \times \frac{9}{9} =$ ■

d. $\frac{3}{5} \times \frac{4}{2} =$ ■

e. $\frac{6}{5} - \frac{1}{5} =$ ■

f. $\frac{14}{19} - \frac{14}{19} =$ ■

Part 10 Work each problem. Remember the unit name.

a. Each bottle holds 12 ounces. There are 5 bottles. How much do they hold in all?

b. A person divides 200 pounds of sand into 4 piles that have the same weight. What is the weight of each pile?

c. A train travels at the steady rate of 30 miles each hour. If the train travels for 6 hours, how many miles does it travel?

Part 11 Write the complete equation for each item.

a. ■ x 7 = 42

b. 3 x 11 = ■

c. ■ – 1 = 10

d. ■ x 4 = 16

e. 20 x 3 = ■

Part 12 Write the equation to show the whole number each fraction equals.

a. $\frac{15}{5}$ b. $\frac{28}{4}$

c. $\frac{9}{9}$ d. $\frac{54}{9}$

Part 13 Copy and work each problem.

a. $2\overline{)870}$ b. $8\overline{)176}$ c. $6\overline{)210}$

d. $\begin{array}{r} 42 \\ \times\ 24 \\ \hline \end{array}$ e. $\begin{array}{r} 18 \\ \times\ 23 \\ \hline \end{array}$ f. $\begin{array}{r} 16 \\ 804 \\ +\ 93 \\ \hline \end{array}$

Part 14 Copy and work each problem.

a. $\frac{2}{3} \times \frac{4}{3} \times \frac{7}{3} =$ ■ b. $\frac{11}{7} - \frac{8}{7} + \frac{2}{7} =$ ■

Part J

e. unshaded 600, shaded 513, bar □ → 600 + 513 = 1113 meters

c. shaded 14, unshaded □, bar 89 → 89 − 14 = 75 tons

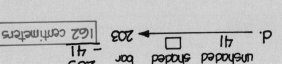

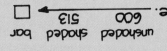

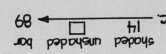

d. unshaded 41, shaded □, bar 203 → 203 − 41 = 162 centimeters

Lesson 9

Part 1

- The bottom number of a fraction is called the **denominator.**
- The denominator of this fraction is 3. $\frac{5}{3}$

Part 2

Rewrite each item with the whole number written as a fraction and work it.

Sample problem $1 + \frac{2}{5} = ■$

$\frac{5}{5} + \frac{2}{5} = \frac{7}{5}$

a. $\frac{12}{7} - 1 = ■$ b. $1 - \frac{3}{4} = ■$ c. $1 + \frac{7}{10} = ■$ d. $\frac{6}{5} + 1 = ■$

Part 3

For each item, write the complete equation. If the fractions shown are equivalent, write a simple equation below.

a. $\frac{2}{7}, \frac{14}{14}$ b. $\frac{4}{3}, \frac{24}{18}$ c. $\frac{7}{4}, \frac{21}{16}$ d. $\frac{2}{10}, \frac{8}{40}$

Part 4

Copy and complete each table.

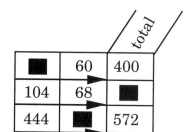

a.

		total
■	60	400
104	68	■
444	■	572

b.

		total
42	278	■
■	138	199
103	■	519

Part 5 For each item, write the fraction equation.

a. $1 + 1 + 1 = 3$

$\frac{\blacksquare}{6} + \frac{\blacksquare}{6} + \frac{\blacksquare}{6} = \blacksquare$

b. $1 + 1 + 1 + 1 = 4$

$\frac{\blacksquare}{4} + \frac{\blacksquare}{4} + \frac{\blacksquare}{4} + \frac{\blacksquare}{4} = \blacksquare$

c. $1 + 1 + 1 + 1 = 4$

$\frac{\blacksquare}{9} + \frac{\blacksquare}{9} + \frac{\blacksquare}{9} + \frac{\blacksquare}{9} = \blacksquare$

Part 6 For each item, make a number family with three names.

> ***Sample sentence 1***
>
> Some of the children were girls.
>
> girls boys children →
>
> ***Sample sentence 2***
>
> Some of the cars were dirty.
>
> dirty not dirty cars →

a. Some of the glasses are full.

b. Some of the children wore shoes.

c. Some of the bricks were hot.

d. Some of the women were not sunburned.

e. Some of the rabbits are sleeping.

Part 7 Copy the table and complete it.

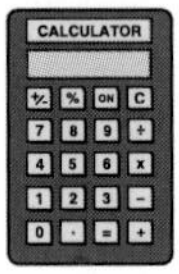

	Multiplication	Division
a.	$11\,(39) = \blacksquare$	$\begin{array}{r}39\\ \overline{)\quad}\end{array}$
b.	$26\,(\blacksquare) = \blacksquare$	$26\overline{)780}$
c.	$24\,(\blacksquare) = \blacksquare$	$24\overline{)696}$
d.	$39\,(17) = \blacksquare$	$\begin{array}{r}17\\ \overline{)\quad}\end{array}$

Independent Work

Part 8 **Copy each problem. Write the missing fraction.**

a. $\frac{3}{4}\left(\blacksquare\right) = \frac{21}{12}$ b. $\frac{6}{5}\left(\frac{3}{4}\right) = \blacksquare$

c. $\frac{2}{7}\left(\blacksquare\right) = \frac{4}{70}$ d. $\frac{4}{5}\left(\frac{1}{2}\right) = \blacksquare$

Part 9 **Copy and work each problem.**

a. $\frac{3}{5} + \frac{9}{5} - \frac{2}{5} = \blacksquare$ b. $\frac{3}{5} \times \frac{9}{5} \times \frac{3}{2} = \blacksquare$

c. $\frac{9}{5} - \frac{8}{5} = \blacksquare$ d. $\frac{2}{10} - \frac{2}{10} = \blacksquare$

Part 10 **Copy the families that you can work and work them.**

a. $\xrightarrow{400 \quad \blacksquare} 560$

b. $\xrightarrow{\blacksquare \quad 112} 99$

c. $\xrightarrow{311 \quad 29} \blacksquare$

d. $\xrightarrow{560 \quad \blacksquare} 305$

e. $\xrightarrow{26 \quad 89} \blacksquare$

f. $\xrightarrow{\blacksquare \quad 48} 97$

Part 11 **Work each problem. Remember the unit name.**

a. A farmer fed his cows 693 pounds of grain. The farmer had 9 cows. If each cow ate the same amount, how many pounds of grain did each cow eat?

b. Each stack of bricks weighed 312 pounds. There were 11 stacks of bricks. What was the total weight for all the bricks?

c. A truck traveled a distance of 200 miles. The truck traveled the same distance each hour. The truck traveled for 5 hours. How many miles did the truck travel each hour?

d. A man divided his fortune equally among his 4 children. If the man's fortune was worth $260,000, how much did each child receive?

e. Basketballs cost $22 each. A team purchased 11 basketballs. How much did they cost?

Part 12 **Write the equation to show the whole number each fraction equals.**

a. $\frac{16}{8}$ b. $\frac{3}{3}$ c. $\frac{42}{7}$

d. $\frac{26}{26}$ e. $\frac{90}{9}$ f. $\frac{51}{1}$

Part 13 **Copy and work each problem.**

a. $2\overline{)2104}$

b. $\begin{array}{r} 864 \\ +\ \ 9 \\ \hline \end{array}$

c. $\begin{array}{r} 357 \\ \times\ 23 \\ \hline \end{array}$

d. $\begin{array}{r} 8030 \\ -\ 6921 \\ \hline \end{array}$

e. $3\overline{)78}$

f. $\begin{array}{r} 1024 \\ 307 \\ +\ 4590 \\ \hline \end{array}$

Lesson 10

Part 1 **For each item, write the complete equation.**

Sample 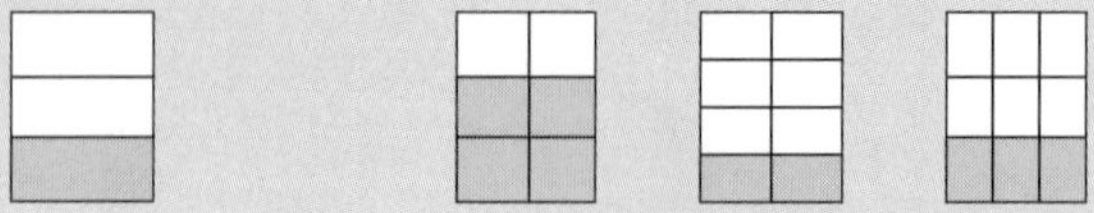

a. 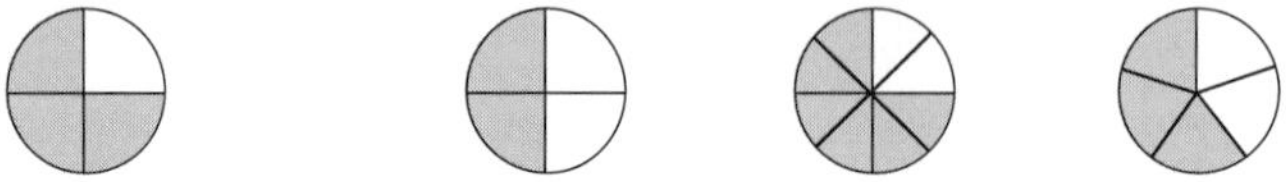

b. 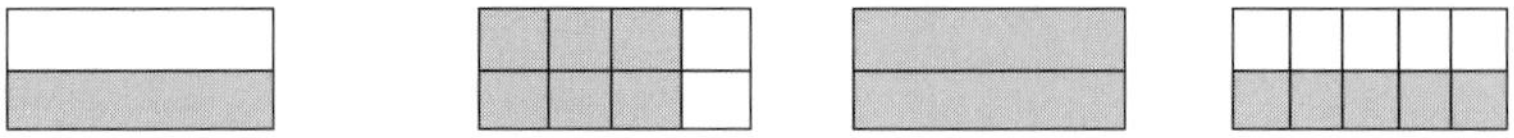

c. 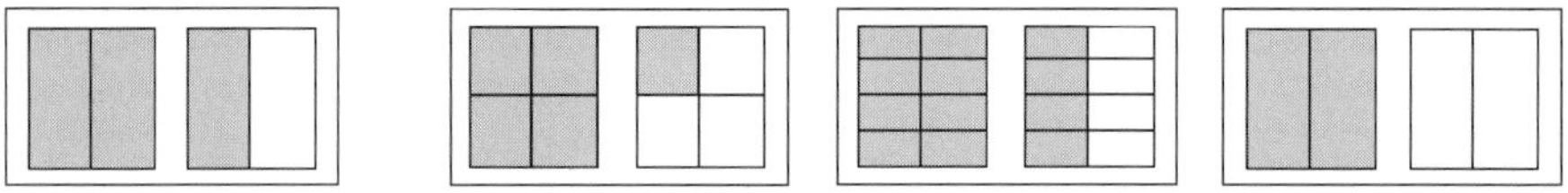

Part 2 **For each item, make a number family with three names.**

a. Some of the roads are not paved.

b. Many of the students passed the test.

c. A few of the doors are open.

d. Lots of dogs were howling.

e. A few boys are sweating.

Part 3 **Rewrite each item with the whole number written as a fraction and work it.**

a. $1 + \frac{7}{20} = ■$

b. $1 - \frac{3}{16} = ■$

c. $\frac{7}{3} + 1 = ■$

d. $\frac{16}{15} - 1 = ■$

Part J

b. $\frac{16}{16} - \frac{3}{16} = \boxed{\frac{13}{16}}$ c. $\frac{7}{3} + \frac{3}{3} = \boxed{\frac{10}{3}}$ d. $\frac{16}{15} - \frac{15}{15} = \boxed{\frac{1}{15}}$

Test 1

Part 1

For each item, write a multiplication equation. If the fractions are equivalent, write a simple equation below.

a. $\frac{7}{4}, \frac{28}{24}$ b. $\frac{2}{3}, \frac{12}{15}$ c. $\frac{6}{8}, \frac{18}{24}$

Part 2

Make a complete number family for each item. Write the answer as a number and unit name. Box the answer.

a. A bar has a shaded part and an unshaded part. The shaded part weighs 46 pounds. The entire bar weighs 111 pounds. How much does the unshaded part weigh?

b. The unshaded part of a bar is 23 inches long. The shaded part is 35 inches long. How long is the entire bar?

Part 3

Copy and work the problems you can work the way they are written. Don't copy any of the problems you can't work.

a. $\frac{4}{8} \times \frac{3}{5} =$ [redacted]

b. $\frac{17}{5} - \frac{5}{5} =$ [redacted]

c. $\frac{2}{7} + \frac{7}{1} =$ [redacted]

d. $\frac{3}{6} \times \frac{7}{6} =$ [redacted]

e. $\frac{1}{9} + \frac{10}{9} =$ [redacted]

f. $\frac{6}{4} - \frac{5}{4} =$ [redacted]

g. $\frac{11}{7} + \frac{11}{11} =$ [redacted]

Part 4 **Copy all the families that are not impossible and figure out the missing numbers.**

a. 784 39 → ■

b. ■ 361 → 294

c. ■ 210 → 530

d. 216 ■ → 106

e. 580 21 → ■

f. 29 ■ → 38

Part 5 **Write the equation to show the whole number each fraction equals.**

a. $\frac{30}{6}$ b. $\frac{12}{4}$ c. $\frac{8}{8}$ d. $\frac{32}{8}$

Part 6 **Write the multiplication or division problem for each word problem. Show the answer as a number and a unit name. Box the answer.**

a. A pie is divided into 7 pieces that are the same weight. Each piece weighs 6 ounces. How much does the entire pie weigh?

b. A loaf of bread has 20 slices that are the same weight. The entire loaf weighs 500 grams. How much does each slice weigh?

Part 7 **Copy and complete the table.**

	Multiplication	Division
a.	8 (■) = 256	
b.	24 (■) = 216	
c.	3 (■) = 369	

Lesson 11

Part 1

- You've learned a name for the bottom number of a fraction. It's the **denominator.**
- The top number of a fraction is called the **numerator.**

Part 2

Write the fraction for each description.

a. The denominator of the fraction is 4. The fraction equals 3 whole units.

b. The denominator of the fraction is 8. The fraction equals 4 whole units.

c. The denominator is 6 and the fraction equals 5.

d. The denominator is 3 and the fraction equals 5 whole units.

e. The denominator is 7. The fraction equals 3.

f. The denominator is 9. The fraction equals 10.

Part 3 Work each item.

Here are the steps:

1. Make a family with three names and two numbers.
2. Calculate the missing number.
3. Write the unit name for the answer and box the answer.

a. There were sick sheep and sheep that were not sick. If there were 68 sheep in all and 44 of them were not sick, how many were sick?

b. There were large stones and small stones. There were 178 stones in all. 89 of them were large. How many of them were small?

c. There were 345 books in a room. 239 of these books were new. How many were used?

d. There were 123 older boys and 74 younger boys on a camping trip. How many boys were there in all?

Part 4

- You can write any division problem as a fraction.
- When you read the division problem, the first number you say is the top number of the fraction. That's the numerator.
- Here's a division problem: $19\overline{)266}$
- When you work division problems on a calculator, the division sign is like a diagram of a fraction.
- 266 is the numerator: [2 6 6]
- The division sign is like a fraction bar: [÷] → $\frac{266}{19}$
- 19 is the denominator: [1 9]
- Remember, when you say the division problem, the first number you say is the numerator of the fraction.

Part 5

For each item, write the fraction. Then complete the equation to show the whole number each fraction equals.

a. $8\overline{)560}$ b. $3\overline{)891}$ c. $25\overline{)1125}$ d. $18\overline{)3780}$

Part 6

Copy each table. Fill in the missing numbers.

a.

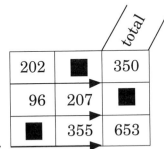

		total
202	■	350
96	207	■
■	355	653

b.

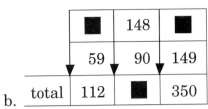

	■	148	■
	59	90	149
total	112	■	350

Part 7

Copy each equation. Write parentheses and the missing fraction that equals 1.

a. $\frac{1}{8} \qquad = \frac{6}{48}$ b. $\frac{7}{9} \qquad = \frac{35}{45}$ c. $\frac{9}{1} \qquad = \frac{63}{7}$

d. $\frac{2}{9} \qquad = \frac{16}{72}$ e. $\frac{3}{4} \qquad = \frac{12}{16}$

Part 8

For each item, write the complete equation.

a.

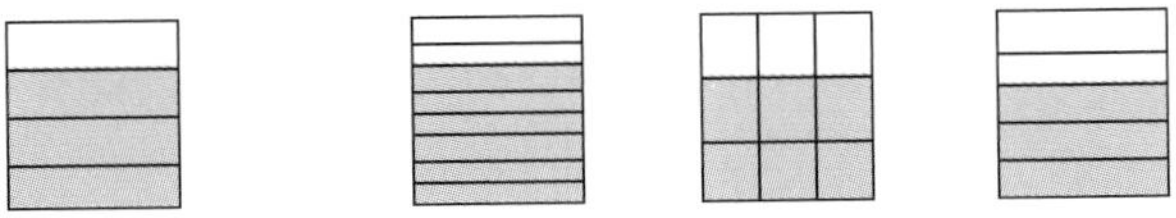

b.

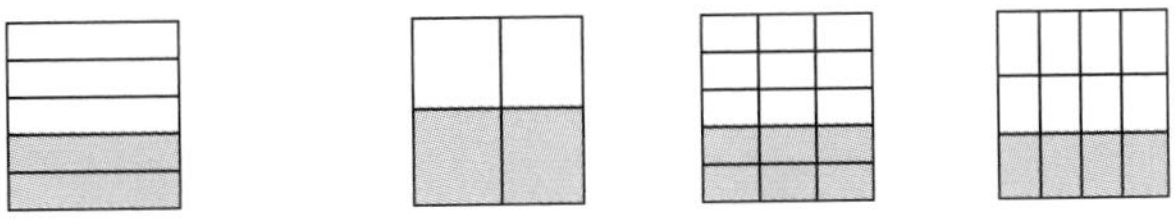

Independent Work

Part 9

Work each problem. Remember the unit name.

a. There were 12 baskets. Each basket contained 4 pounds of cheese. How many pounds of cheese were there in all?

b. A parcel of land was divided into 8 smaller parcels, each the same size. The whole parcel contained 560 acres. How many acres were in each smaller parcel?

c. It cost a farmer \$8 to vaccinate each cow. The farmer spent \$96 to vaccinate all of his cows. How many cows did the farmer have?

Part 10

For each item, write a multiplication equation. If the fractions shown are equivalent, write the simple equation below.

a. $\frac{7}{3}, \frac{28}{24}$ b. $\frac{5}{9}, \frac{35}{63}$

c. $\frac{2}{1}, \frac{18}{9}$ d. $\frac{10}{7}, \frac{100}{49}$

Part 11

Write the complete equation for each item.

a. $26 + \blacksquare = 29$ d. $\blacksquare - 10 = 0$ f. $\blacksquare + 3 = 11$

b. $\blacksquare \times 9 = 36$ e. $17 - \blacksquare = 8$ g. $\blacksquare \times 6 = 42$

c. $7 \times 7 = \blacksquare$

Part 12

For each item, write an equation that shows the fraction and the whole number it equals.

a. $\frac{12}{4}$ b. $\frac{12}{6}$ c. $\frac{12}{3}$ d. $\frac{12}{12}$ e. $\frac{8}{4}$ f. $\frac{8}{8}$ g. $\frac{8}{2}$

Part 13

Copy the table and complete all the rows. You can use your calculator.

	Multiplication	Division
a.	$7 \times \blacksquare = 728$	
b.		$4\overline{)224}$
c.	$2 \times \blacksquare = 80$	
d.	$3 \times \blacksquare = 87$	

Part 14

Copy and work each problem.

a. $\frac{3}{8} \times \frac{1}{8} = \blacksquare$

b. $\frac{4}{8} - \frac{3}{8} - \frac{1}{8} = \blacksquare$

c. $\frac{2}{5} \times \frac{4}{2} \times \frac{1}{3} = \blacksquare$

d. $\frac{12}{10} + \frac{5}{10} = \blacksquare$

Lesson 12

Part 1

- You've worked multiplication problems that have a missing fraction.
- You work the problem for the numerators and the problem for the denominators.
- Here's a problem: $\frac{3}{2}\left(\blacksquare\right) = \frac{12}{10}$
- Here's the complete equation: $\frac{3}{2}\left(\frac{4}{5}\right) = \frac{12}{10}$
- Here's a different kind of problem: $\frac{4}{7} = \frac{12}{\square}$
- Part of the last fraction is missing. The equal sign tells you that the two fractions are equivalent. You multiply the first fraction by 1 to get the other fraction. $\frac{4}{7}\left(\blacksquare\right) = \frac{12}{\square}$
- To figure out the fraction that equals 1, you work the problem for either the numerators or the denominators.
- You can't work the problem for the denominators. You don't have enough numbers. $\frac{4}{7}\left(\frac{\blacksquare}{\blacksquare}\right) = \frac{12}{\square}$
- But you can work the problem for the numerators. $\frac{4}{7}\left(\frac{\blacksquare}{\blacksquare}\right) = \frac{12}{\square}$
- The missing value is 3. So the fraction that equals 1 is $\frac{3}{3}$. $\frac{4}{7}\left(\frac{3}{3}\right) = \frac{12}{\square}$
- Now you can work the problem for the denominators. $\frac{4}{7}\left(\frac{3}{3}\right) = \frac{12}{21}$
- Remember the steps:
  1. Work the problem for either the numerators or the denominators.
  2. Complete the fraction that equals 1.
  3. Work the problem for the number that goes in the box.

Part 2 Complete each pair of equivalent fractions.

Sample problem

$\frac{10}{2} = \frac{■}{18}$

a. $\frac{7}{3} = \frac{35}{■}$ b. $\frac{2}{5} = \frac{12}{■}$ c. $\frac{4}{9} = \frac{■}{54}$

Part 3 Rewrite each equation so the value that is alone comes first.

Sample equation

12 + 15 = 27

a. 115 – 8 = 107

b. 2 x 7 = 14

c. 96 + 112 = 208

d. 9 x 17 = 153

Part 4 Copy the table and complete it.

	Division	Fraction equation
Sample	**$77\overline{)462}$ = 6**	**$\frac{462}{77} = 6$**
a.	$15\overline{)495}$	
b.	$61\overline{)793}$	
c.	$13\overline{)585}$	

Part 5 For each item, make a number family with names. Then figure out the answer to the question.

a. There were 567 adults at a picnic. The rest of the people were children. There were 888 people in all. How many children were at the picnic?

b. There were 480 cars on a lot. 89 of those cars were red. How many cars were not red?

c. Men and women worked in a large office. 128 were women. 148 were men. How many people worked in the office?

Part 6 **Write the fraction for each description.**

a. The fraction has a denominator of 2. The fraction equals 13.
b. The fraction has a denominator of 8. The fraction equals 6.
c. The fraction equals 6. The fraction has a denominator of 1.
d. The fraction equals 6 and has a denominator of 5.

Part 7

- Some problems tell about **ratios.** When you work ratio problems, you have **two unit names.** You write fractions so that one unit name tells about the numerator and the other unit name tells about the denominator.
- Here's part of a problem:

  In a mixture, there are 3 parts of sand for every 5 parts of water.

- The unit names are **parts of sand** and **parts of water.**
- **Parts of sand** is mentioned first, so you write it as the top name for your fraction. **Parts of water** goes on the bottom.

  $\frac{\textbf{parts of sand}}{\textbf{parts of water}}$

- Now you put the numbers where they go.
- There are 3 parts of sand.

  $\frac{\textbf{parts of sand}}{\textbf{parts of water}} \quad \frac{3}{}$

- There are 5 parts of water.

  $\frac{\textbf{parts of sand}}{\textbf{parts of water}} \quad \frac{3}{5}$

Part 8 **For each item, make a fraction with unit names and a fraction with numbers.**

a. There are 8 dogs for every 300 fleas.
b. There are 7 pounds in every 3 cartons.
c. There were 100 plates for every 13 pots.
d. Every 4 containers hold 11 gallons.
e. There was a ratio of 9 children to every 5 adults.

Part 9

Copy the table. Figure out the missing numbers and write them in the table.

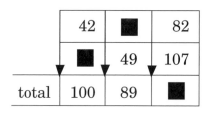

	42	$\blacksquare$	82
	$\blacksquare$	49	107
total	100	89	$\blacksquare$

Independent Work

Part 10

Write an equation to show the first fraction and the equivalent fraction. Show the fraction that equals 1. Use this form: $\blacksquare(\ ) = \blacksquare$

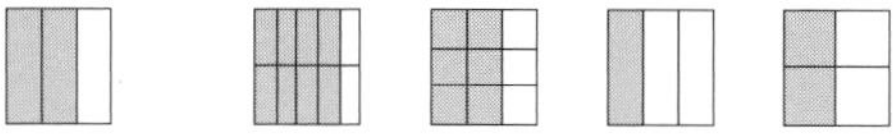

Part 11

Copy only the problems you can work the way they are written and work them.

a. $\frac{7}{4} + \frac{7}{11} = \blacksquare$

b. $\frac{10}{1} \times \frac{3}{5} = \blacksquare$

c. $\frac{4}{9} \times \frac{1}{9} = \blacksquare$

d. $\frac{10}{5} - \frac{9}{5} = \blacksquare$

e. $\frac{9}{32} + \frac{0}{32} = \blacksquare$

f. $\frac{10}{1} - \frac{10}{7} = \blacksquare$

g. $\frac{5}{11} - \frac{5}{11} = \blacksquare$

h. $\frac{3}{2} \times \frac{1}{2} = \blacksquare$

i. $\frac{16}{10} + \frac{10}{4} = \blacksquare$

Part 12

Write the complete equation for each item.

a. $5 \times \blacksquare = 0$

b. $\blacksquare + 1 = 20$

c. $\blacksquare \times 3 = 15$

d. $\blacksquare - 3 = 15$

e. $\blacksquare \times 12 = 12$

f. $\blacksquare + 12 = 12$

Part 13

Rewrite each item with the whole number written as a fraction and work it.

a. $1 - \frac{4}{9} = \blacksquare$

b. $\frac{2}{5} + 1 = \blacksquare$

c. $\frac{4}{2} - 1 = \blacksquare$

d. $\frac{8}{3} - 1 = \blacksquare$

e. $1 + \frac{10}{3} = \blacksquare$

Part 14

Copy each problem and work it.

a. $\frac{3}{7}(\blacksquare) = \frac{21}{21}$

b. $\frac{4}{1}\left(\frac{7}{35}\right) = \blacksquare$

c. $\frac{10}{3}(\blacksquare) = \frac{10}{30}$

d. $\frac{8}{19}\left(\frac{0}{5}\right) = \blacksquare$

Part 15 **Copy the table and complete all the rows.**

	Multiplication	Division
a.	7 x ■ = 945	
b.		6 ⟌ 414
c.	5 x ■ = 85	

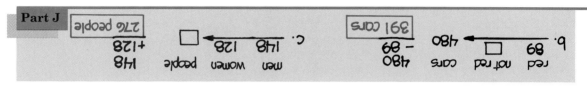

Part J

b. red 89, not red □, cars 480 → 480 − 89 = 391 cars

c. men 148, women 128 → people □; 148 + 128 = 276 people

I don't think that's the way you're supposed to box your answer.
4,1 6$\not{5}\not{0}$ − 148
502 red books

Lesson 13

Part 1 **Complete each pair of equivalent fractions.**

a. $\frac{4}{5}\left(\frac{8}{8}\right) = \blacksquare$

b. $\frac{4}{5} = \frac{28}{\blacksquare}$

c. $\frac{10}{9} = \frac{\blacksquare}{18}$

d. $\frac{7}{3}\left(\frac{10}{10}\right) = \blacksquare$

Part 2 **Rewrite each equation so it begins with the value that is alone on one side of the equation.**

a. 900 – 500 = 400

b. 11 x 9 = 99

c. 358 + 700 = 1058

d. 53 x 40 = 2120

Part 3 **For each item, write the names and the fraction.**

a. 15 square feet of cardboard are used for every 4 boxes.

b. The truck moved at the steady rate of 85 miles every 3 hours.

c. The classroom needed 5 books for every 9 students.

Part 4 **For each item, write the names and the fraction.**

a. The ratio of dogs to cats is 2 to 7.

b. The ratio of cats to fleas is 4 to 81.

c. The ratio of doors to windows was 1 to 4.

Part 5

- The perimeter is the distance around a figure. To find the perimeter, you add the length of each side.

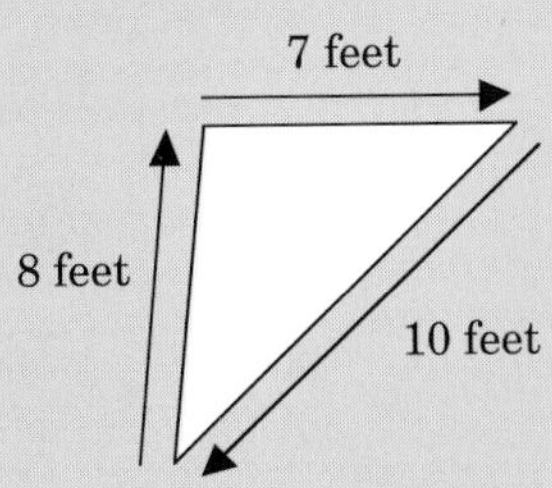

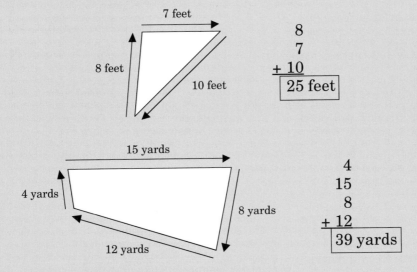

$$\begin{array}{r} 8 \\ 7 \\ +\ 10 \\ \hline \boxed{25 \text{ feet}} \end{array}$$

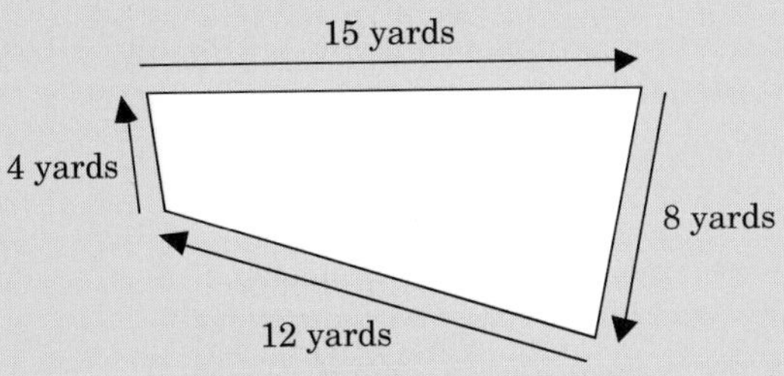

$$\begin{array}{r} 4 \\ 15 \\ 8 \\ +\ 12 \\ \hline \boxed{39 \text{ yards}} \end{array}$$

- Perimeters have a number and a unit name.

Part 6 **Find the perimeter of each figure. Remember the unit name.**

a.

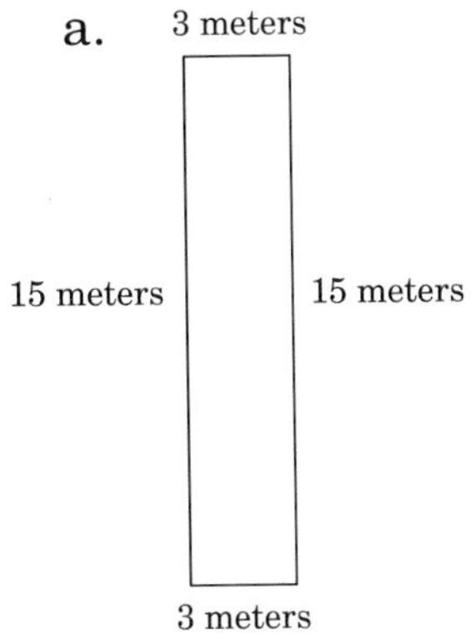

b.

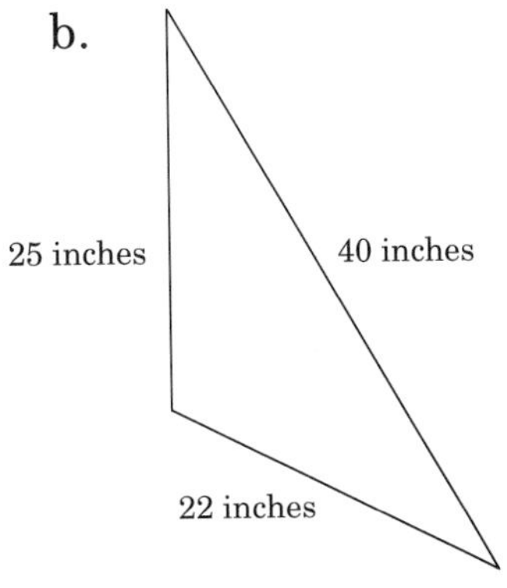

c.

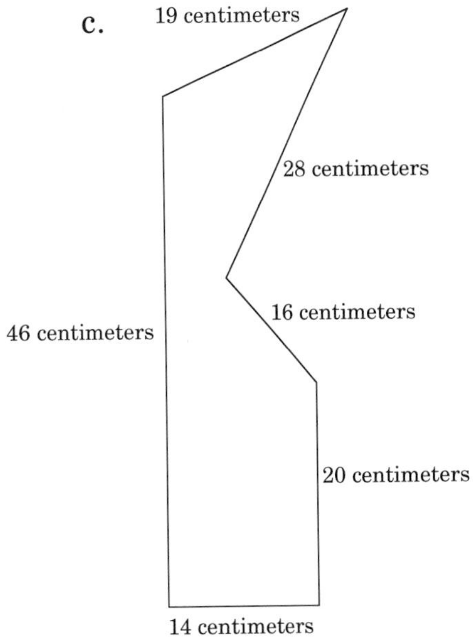

Part 7 Copy the table. Fill in the missing numbers.

- If there are two numbers in a row, you can figure out the missing number in that row.
- If there's only one number in a row, you can't figure out the missing numbers.
- The same rule holds for columns.

			total
	150	■	245
	■	98	■
total	172	193	■

Part 8 Copy the table and complete it.

	Multiplication	Division	Fraction equation
Sample	**12 (96) = 1152**	$\begin{array}{r}96\\12\overline{)1152}\end{array}$	$\frac{1152}{12} = 96$
a.	37 (■) = 925	$\overline{)}$	
b.	52 (■) = 676		

Independent Work

Part 9 For each item, make a number family with names. Then figure out the answer to the question.

a. In the forest, 64 trees were dead. There were 158 trees in the forest. How many of the trees in the forest were alive?

b. In March, 11 days were rainy and 20 days were not rainy. How many days were there altogether in March?

c. John had baseball cards and football cards. He had 390 cards in all. He had 173 football cards. How many baseball cards did John have?

Part 10 For each fraction that equals a whole number, write the equation to show the fraction and the whole number it equals.

a. $\frac{2}{1}$ b. $\frac{20}{20}$ c. $\frac{4}{12}$ d. $\frac{40}{39}$ e. $\frac{12}{4}$ f. $\frac{56}{57}$ g. $\frac{1}{2}$ h. $\frac{3}{9}$ i. $\frac{14}{7}$

Part 11 **Copy and complete each number family that is not impossible.**

a. 130 ■ → 121

b. ■ 59 → 210

c. 130 121 → ■

d. ■ 147 → 129

e. 580 ■ → 760

f. 68 ■ → 88

Part 12 **For each item write a multiplication equation. If the fractions shown are equivalent, write the simple equation below.**

a. $\frac{9}{4}$, $\frac{36}{12}$

b. $\frac{2}{5}$, $\frac{42}{105}$

c. $\frac{7}{8}$, $\frac{63}{72}$

d. $\frac{10}{3}$, $\frac{30}{15}$

Part 13 **Work each item. Remember the unit name.**

a. The grass grew 3 millimeters each day. The grass grew for 17 days. How much did it grow?

b. A snail moved a total distance of 36 inches. If the snail moved 4 inches each minute, for how many minutes did the snail move?

c. James divided his baseball cards into 4 equal-sized stacks. There were 20 cards in each stack. How many cards did James have in all?

Part 14 **For each item, write the complete equation.**

a. ■ + 2 = 10

b. 12 x 4 = ■

c. ■ x 6 = 24

d. 20 x ■ = 60

e. 56 + ■ = 60

f. ■ + 11 = 13

g. 8 x 56 = ■

Part 15 **Copy and work each item.**

a. 36 x 14

b. 3)483

c. 4)260

d. 23 x 45

Lesson 14

Part 1 **For each item, write the complete equation.**

Sample problem

$3 = \frac{\blacksquare}{5}$

a. $5 = \frac{\blacksquare}{8}$

b. $\frac{\blacksquare}{7} = 6$

c. $9 = \frac{\blacksquare}{4}$

d. $\frac{\blacksquare}{10} = 3$

e. $10 = \frac{\blacksquare}{9}$

Part 2

- You've worked with equations that have an unknown after the equal sign. — $50 - 7 = \blacksquare$
- We'll write the same equation so the unknown comes first. The side with 50 – 7 comes after the equal sign. — $\square = 50 - 7$
- You figure out the answer the same way you would if the box were on the other side. You figure out what 50 – 7 equals. That's the number that goes in the box. — $\boxed{43} = 50 - 7$

Part 3 **Rewrite each equation so that the box comes first. Then work each problem.**

a. $5 \times 4 = \blacksquare$

b. $13 - 11 = \blacksquare$

c. $50 + 70 = \blacksquare$

d. $30 - 12 = \blacksquare$

Part 4 Write an equation for each item.

a. 5 bottles weigh 2 pounds. How many bottles weigh 20 pounds? $\frac{\text{bottles}}{\text{pounds}} \frac{5}{2} =$ ■

b. 5 bottles weigh 2 pounds. How many pounds do 15 bottles weigh?

c. The ratio of bottles to pounds is 5 to 2. How many pounds do 30 bottles weigh?

d. 5 bottles weigh 2 pounds. How many bottles weigh 30 pounds?

Part 5 Find the perimeter of each figure.

a.

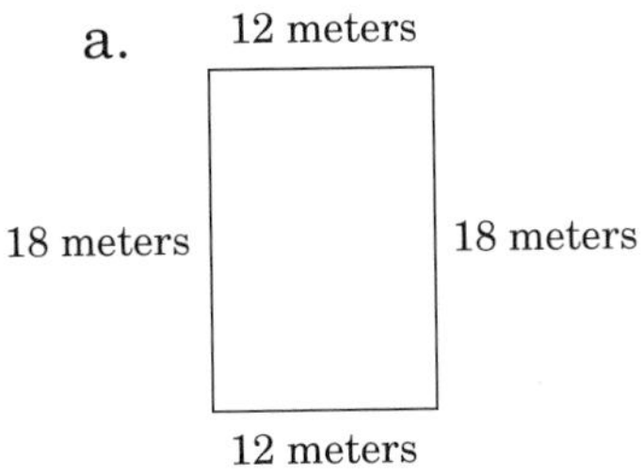

b.

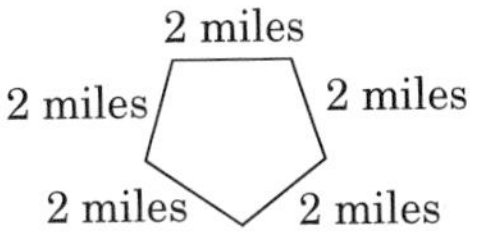

c.

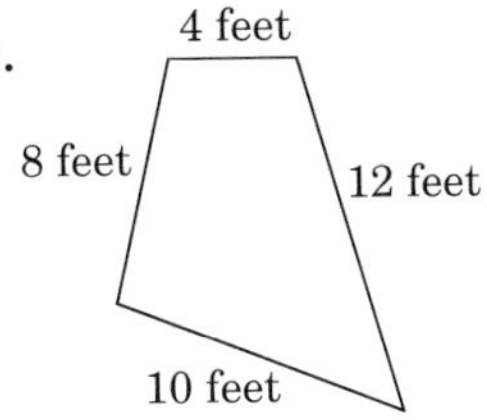

d.

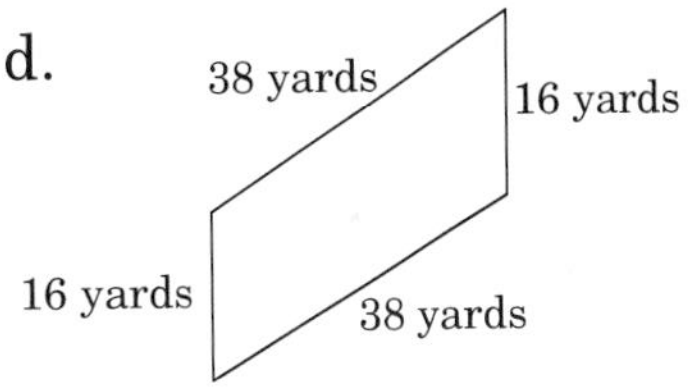

Part 6 Copy each table. Fill in the missing numbers.

a.

			total
	152	■	371
	■	45	■
total	268	264	■

b.

			total
	■	267	381
	172	■	407
total	■	502	■

Part 7 For each item, write the equation to show the whole number the fraction equals.

a. $\frac{2444}{52}$ b. $\frac{282}{47}$ c. $\frac{532}{28}$ d. $\frac{1848}{56}$

Independent Work

Part 8 Write the fraction for each picture.

a. 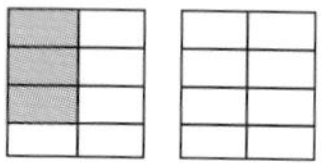

b. 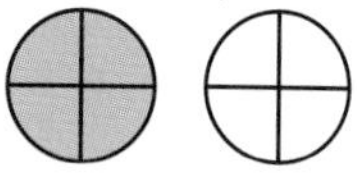

c. 

Part 9 Copy and work all the problems that can be worked the way they are written.

a. $\frac{7}{3} \times \frac{1}{5} =$ ■

b. $\frac{2}{5} + \frac{7}{3} =$ ■

c. $\frac{2}{8} - \frac{1}{8} =$ ■

d. $\frac{5}{4} \times \frac{4}{5} =$ ■

e. $\frac{17}{5} - \frac{10}{5} =$ ■

f. $\frac{8}{3} - \frac{3}{8} =$ ■

g. $\frac{5}{20} - \frac{5}{20} =$ ■

Part 10 Make a number family for each problem. Answer the questions.

a. 45 of the cups were broken. There were 568 cups. How many were not broken?

b. There were 55 melons. 23 were hard. The rest were soft. How many were soft?

c. There were 25 dogs that were howling and 46 dogs that were not howling. How many dogs were there in all?

Part 11 Copy and complete each equation.

a. $\frac{4}{5} = \frac{32}{■}$

b. $\frac{3}{8}\left(\frac{4}{4}\right) =$ ■

c. $\frac{1}{5}\left(\frac{16}{16}\right) =$ ■

d. $\frac{7}{3} = \frac{28}{■}$

e. $\frac{2}{9} = \frac{■}{45}$

Part 12 Copy and complete each equation to show the whole number that equals each fraction.

a. $\frac{20}{10} =$ ■ b. $\frac{4}{1} =$ ■ c. $\frac{35}{7} =$ ■

d. $\frac{56}{7} =$ ■ e. $\frac{28}{7} =$ ■

Part 13 Copy and work each problem.

a. 37 + ■ = 40

b. ■ x 26 = 26

c. 7 x ■ = 0

d. 15 – ■ = 10

e. 84 + ■ = 84

f. ■ x 4 = 32

g. 3 x 6 = ■

h. ■ x 1 = 15

i. 19 – ■ = 19

j. 36 – 20 = ■

Part J

a. 18
18
12
+ 12
60 meters

b. 2
2
2
2
+ 2
10 miles

c. 8
4
12
+ 10
34 feet

d. 16
16
38
+ 38
108 yards

Part K

			total
	114	267	381
	172	235	407
b. total	286	502	788

When I think about finding the perimeter of that place, I think about walking all the way around the outside of it.
Why walk around it when we can stop at that empty table and get something to eat?
RESTAURANT

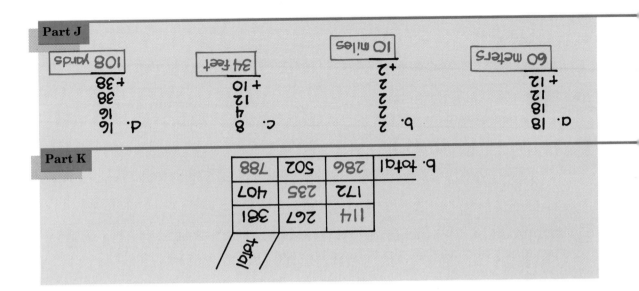

Lesson 15

Part 1 **For each item, write the complete ratio equation and answer the question.**

a. During part of last summer, there were 8 sunny days for every 7 cloudy days. There were 56 sunny days. How many cloudy days were there?

$$\frac{\text{sunny days}}{\text{cloudy days}} \quad \frac{8}{7} \quad = \frac{}{}$$

b. An ant can move 9 inches every 7 seconds. How many seconds would it take the ant to move 639 inches?

$$\frac{\text{inches}}{\text{seconds}} \quad \frac{9}{7} \quad = \frac{}{}$$

c. The ratio of flour to sugar is 2 to 7. How many cups of flour are needed for 28 cups of sugar?

$$\frac{\text{cups of flour}}{\text{cups of sugar}} \quad \frac{2}{7} \quad = \frac{}{}$$

d. Every 3 hours a shop fixes 8 cars. How many hours does it take the shop to fix 48 cars?

$$\frac{\text{hours}}{\text{cars}} \quad \frac{3}{8} \quad = \frac{}{}$$

Part 2 **Rewrite each equation so the box comes first. Then work each problem.**

a. 456 – 49 = ■

b. 13 x 9 = ■

c. 562 + 88 = ■

d. 501 – 275 = ■

Part 3

- Some fractions that are more than 1 do not equal whole numbers. These fractions can be written as mixed numbers.
- The mixed number shows the number of **whole units** and shows the number of **parts that are leftover.** The fraction in the mixed number is always less than 1 whole unit.

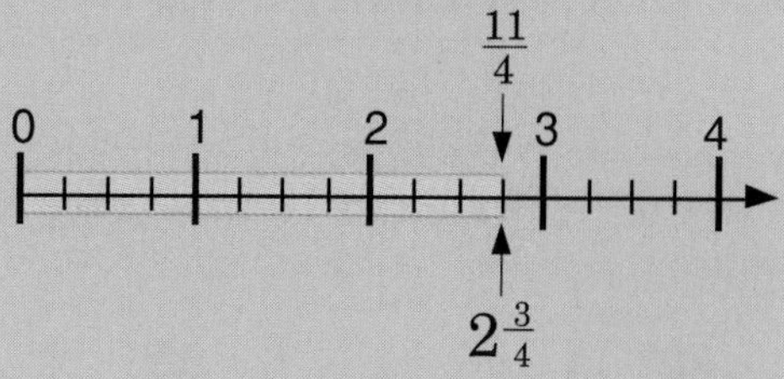

- Here's a bar on the number line:
- The fraction for the whole bar is $\frac{11}{4}$. Each unit is divided into 4 parts. 11 parts are shaded.
- The mixed number that equals $\frac{11}{4}$ is $2\frac{3}{4}$. That's 2 full units and $\frac{3}{4}$ of the next unit.

$$\frac{11}{4} = 2\frac{3}{4}$$

Part 4

For each item, write the fraction. Then complete the equation to show the mixed number the fraction equals.

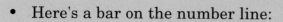

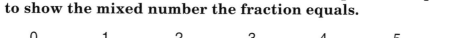

a. 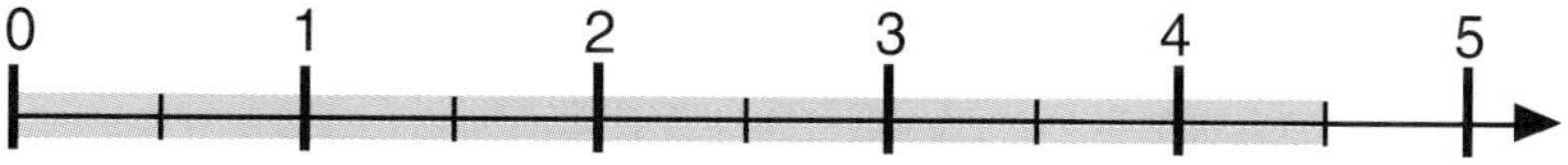

b. 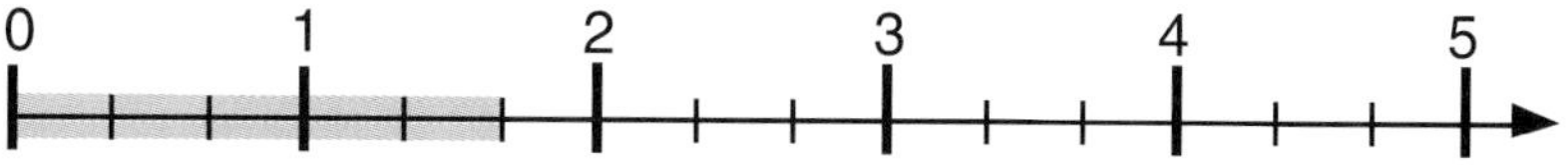

c. 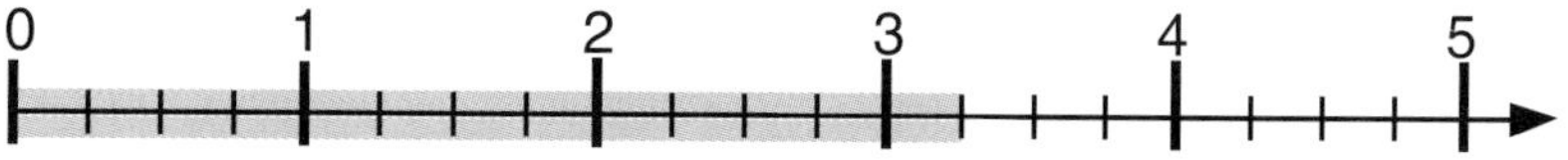

d. 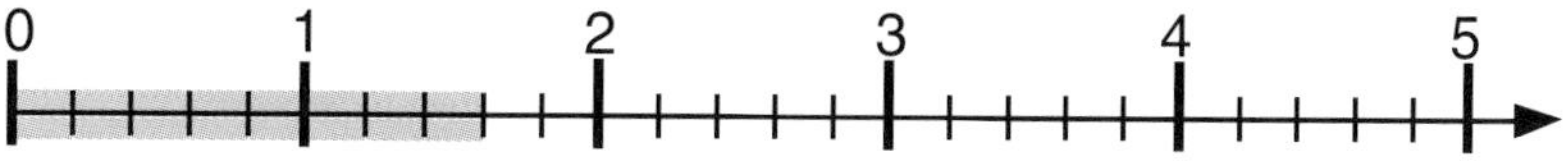

Part 5 Copy and complete each table.

a.

			total
	■	149	383
	■	■	■
total	470	319	■

b.

			total
	88	■	132
	77	66	■
total	■	■	■

Part 6 For each item, write the complete equation.

a. $\frac{\blacksquare}{5} = 4$

b. $\frac{\blacksquare}{4} = 5$

c. $6 = \frac{\blacksquare}{9}$

d. $\frac{\blacksquare}{20} = 3$

Part 7 Find the perimeter of each figure.

a.

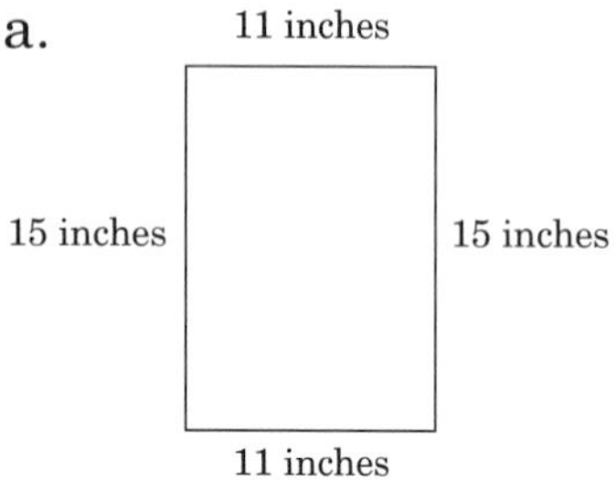

b.

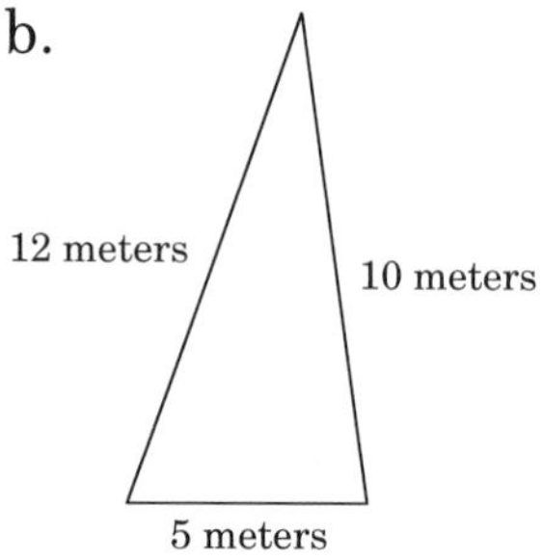

c.

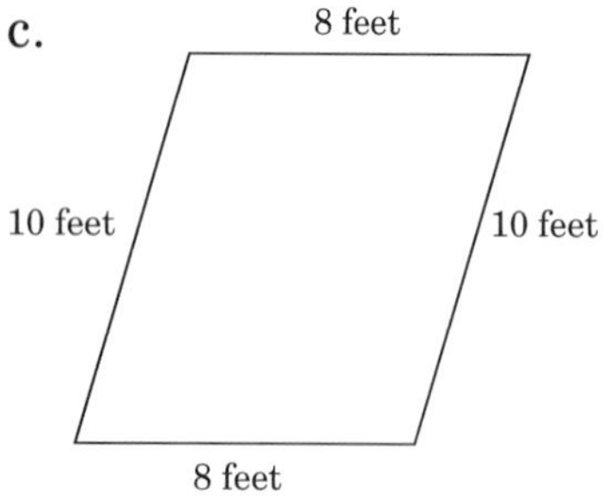

d.

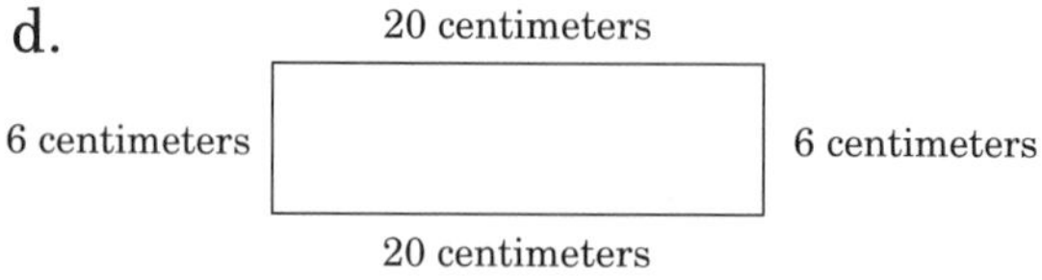

Independent Work

Part 8

Make a number family for each problem. Answer the question.

a. There were 680 windows in a building. 240 windows were clean. The rest were dirty. How many were dirty?

b. A grove had tall trees and trees that were not tall. 160 trees were tall. 167 trees were not tall. How many trees were in the grove?

c. A store had 41 sweaters. 25 of those sweaters had buttons. How many sweaters did not have buttons?

d. In a kennel, 31 dogs had long hair. The rest had short hair. There were 105 dogs in the kennel. How many had short hair?

Part 9

Rewrite each item with the whole number written as a fraction and work it.

a. $1 - \frac{3}{10} =$ ■ b. $\frac{12}{3} - 1 =$ ■

c. $\frac{3}{12} + 1 =$ ■ d. $\frac{5}{5} - 1 =$ ■

Part 10

Copy and complete each equation.

a. $\frac{1035}{23} =$ ■ b. $\frac{250}{2} =$ ■

Part 11

Copy the table and complete all the rows.

	Multiplication	Division	Fraction equation
a.	9 x ■ = 639		
b.	7 x ■ = 168		
c.	2 x ■ = 58		
d.			$\frac{225}{5}$

Part 12

Copy each item and work it.

a. $\frac{5}{11} + \frac{2}{11} =$ ■

b. $\frac{5}{10} - \frac{4}{10} + \frac{2}{10} =$ ■

c. 5 x ■ = 30

d. $\frac{1}{4} \times \frac{3}{2} \times \frac{5}{3} =$ ■

e. ■ – 4 = 6

f. 16 + ■ = 20

g. $\frac{7}{9} \times \frac{3}{9} =$ ■

h. ■ x 3 = 21

i. ■ – 1 = 21

Part J

a. $\frac{20}{5} = 4$ b. $\frac{20}{4} = 5$ c. $6 = \frac{54}{9}$ d. $\frac{60}{20} = 3$

Lesson 16

Part 1

- You can make number families that compare two things. A comparison number family shows the **bigger thing** as the **big number.** The **smaller thing** is one of the **small numbers.** The **difference** between the two things is the other **small number.** It's always the first small number.
- Here are two boards, N and B:
- The dotted part next to board N is the difference.
- B is larger, so it is the name for the big number.
- N is smaller; it's the name for one of the small numbers.
- The name for the other small number is the **difference**.

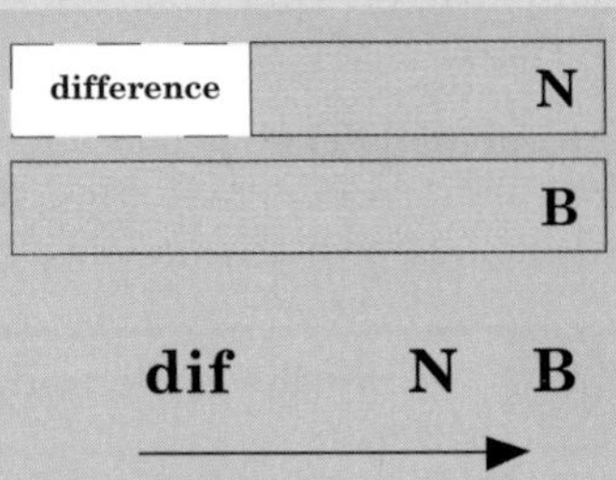

- The difference shows how much you'd have to subtract from B to make it the same size as N.

Part 2

For each item, make a number family with three names.

a.

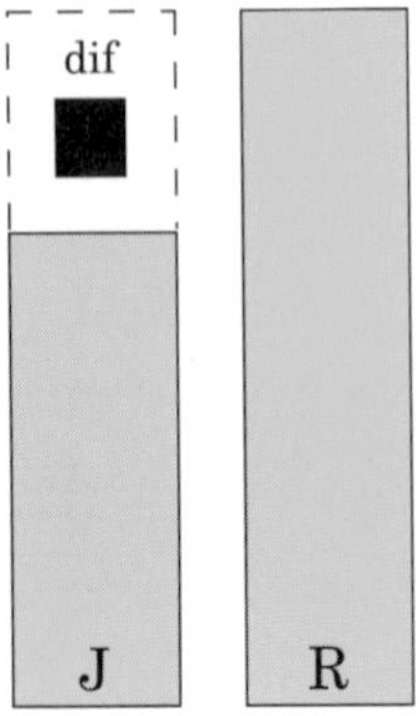

b.

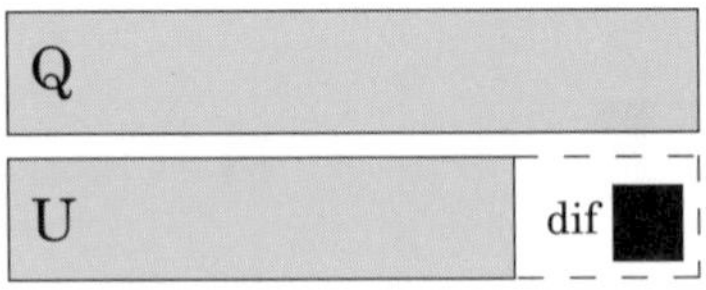

c.

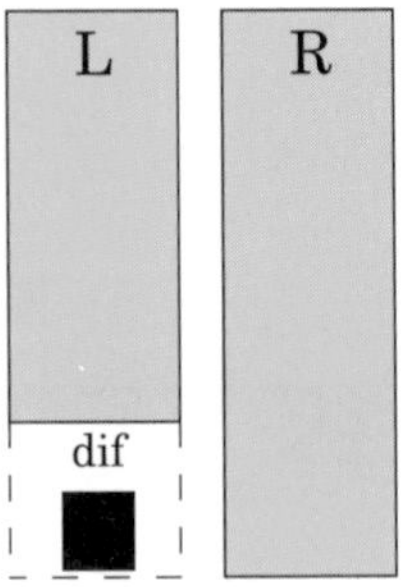

Part 3

For each item, make a number family with three names and two numbers. Then figure out the missing number.

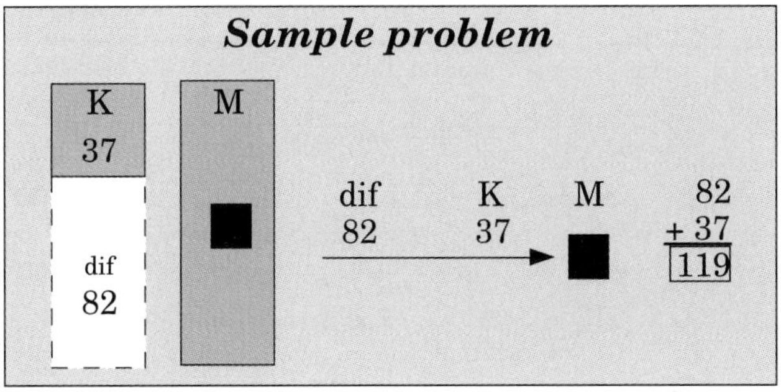

a.

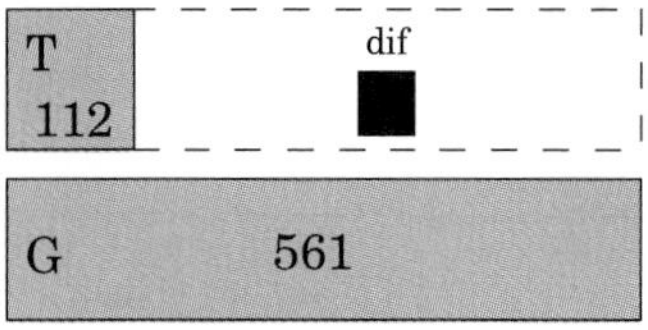

b.

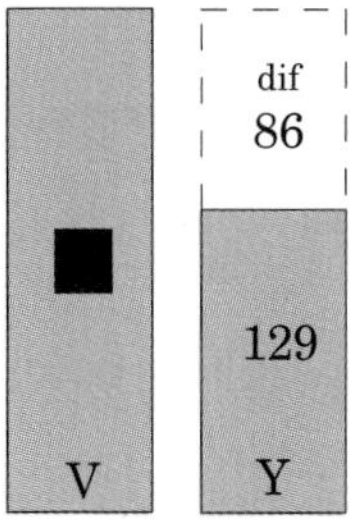

Part 4

For each item, write an equation to show the fraction and the mixed number it equals.

a.

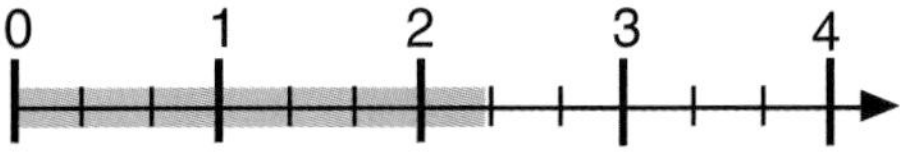

b.

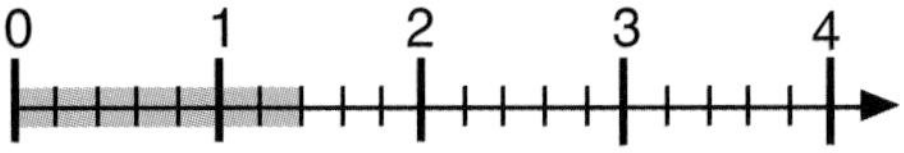

c.

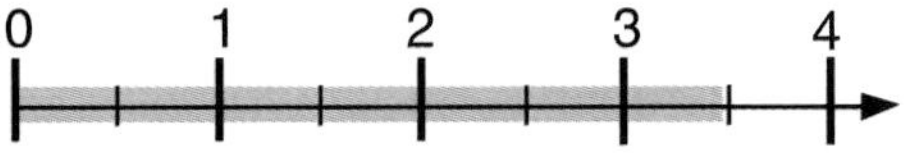

d.

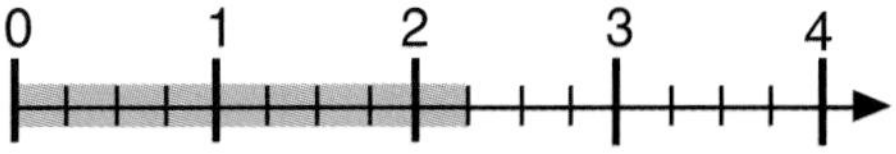

e.

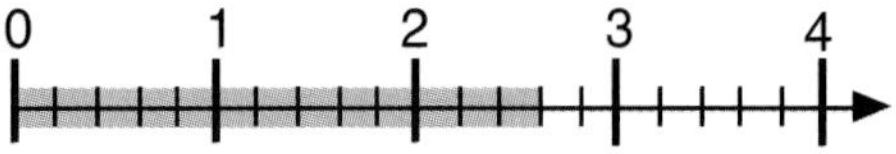

Part 5

Make a ratio equation for each problem. Answer each question.

a. There are 21 fleas for every 3 cats. If there are 21 cats, how many fleas are there?

b. The ratio of trees to cones is 4 to 7. There are 20 trees. How many cones are there?

c. There are 9 buses for every 4 trucks. If there are 72 buses, how many trucks are there?

Part 6

For each item, figure out what T equals. Write a simple equation for T.

> ***Sample problem***
>
> 543 – 345 = T
>
> T = 543 – 345
>
> T = 198

a. 978 + 269 = T

b. 120 x 8 = T

c. 470 – 248 = T

Part 7

- Here's how to write mixed numbers as fractions. $5\frac{2}{3} =$
- First, you copy the denominator. The denominator in the mixed number is 3. $5\frac{2}{3} = \frac{}{3}$
- Next, you figure out the numerator of the fraction. You start with the denominator of the fraction and **multiply** by the whole number. Then you **add** the numerator of the fraction in the mixed number.

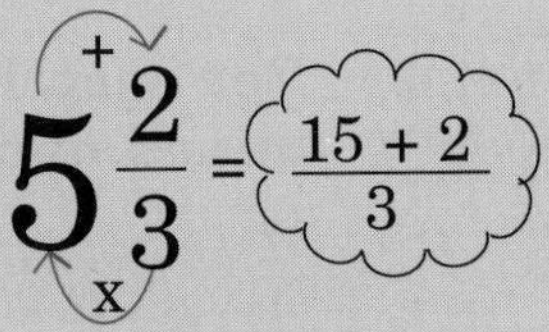

- So, the fraction that equals $5\frac{2}{3}$ is $\frac{17}{3}$. $5\frac{2}{3} = \frac{17}{3}$
- Remember, first **multiply.** Then **add.**

Sample problem $3\frac{1}{4} =$ [redacted] / [redacted]

Part 8

Write each mixed number and the fraction it equals.

a. $6\frac{1}{3} =$ [redacted]

b. $2\frac{4}{5} =$ [redacted]

c. $3\frac{2}{7} =$ [redacted]

Independent Work

Part 9

Copy and complete the table.

			total
	200	■	285
	100	300	■
total	■	385	■

Part 10

Copy each problem and work it.

a. $\blacksquare \times 9 = 54$

d. $16 \times \blacksquare = 0$

g. $\blacksquare \times 11 = 88$

b. $\frac{12}{19} + \frac{3}{19} = \blacksquare$

e. $\frac{3}{5} + \frac{10}{5} = \blacksquare$

h. $\frac{42}{3} - \frac{20}{3} = \blacksquare$

c. $\frac{42}{3} \times \frac{1}{3} = \blacksquare$

f. $\frac{11}{3} - \frac{11}{3} = \blacksquare$

i. $\blacksquare - 197 = 0$

Part 11

Make a number family for each item. Answer the question.

a. 56 of the students were girls. The rest were boys. If there were 111 students in all, how many were boys?

b. A parcel of land was divided into two pieces. The larger piece was 145 acres. The smaller piece was 126 acres. How many acres were in the entire parcel?

c. Janet and Fran collected rocks. Together, the girls had 134 rocks. Janet collected 88 of them. How many rocks did Fran collect?

d. During the first half of January, 21 centimeters of snow fell. During the second half, 33 centimeters of snow fell. What was the total snowfall for the month?

Part 12

Rewrite each item with the whole number written as a fraction and work it.

a. $\frac{7}{5} - 1 = \blacksquare$

b. $1 + \frac{1}{10} = \blacksquare$

c. $1 + \frac{5}{6} = \blacksquare$

d. $1 - \frac{11}{11} = \blacksquare$

Part 13

Write an equation for each pair of fractions. If the fractions shown are equivalent, write the simple equation below.

a. $\frac{3}{2}, \frac{42}{32}$

b. $\frac{7}{9}, \frac{42}{54}$

c. $\frac{4}{5}, \frac{100}{120}$

Part 14

Find the perimeter of each figure.

a.
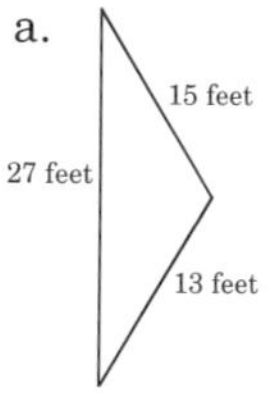

b.
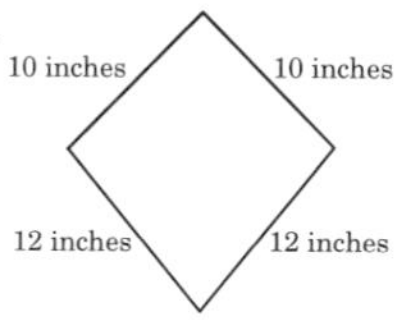

c.
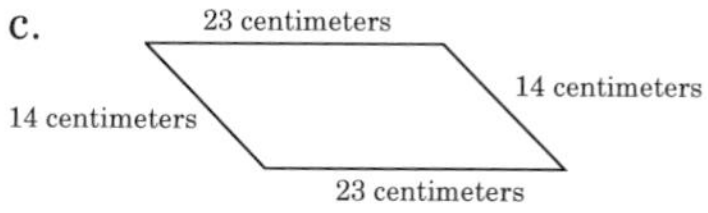

d.
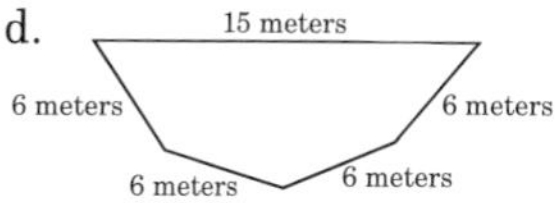

Part 15

For each problem, write the complete equation showing the fraction and the whole number it equals.

a. The denominator of the fraction is 4. The fraction equals 3 whole units.

b. The denominator of the fraction is 8. The fraction equals 10.

c. $\frac{20}{10} = \blacksquare$

d. $\frac{12}{3} = \blacksquare$

e. $\frac{\blacksquare}{3} = 12$

f. $8 = \frac{\blacksquare}{2}$

Lesson 17

Part 1

- You've found the perimeter of rectangles. You can also find the area.
- To find the area of rectangles, you multiply the **base** of the rectangle by the **height.**

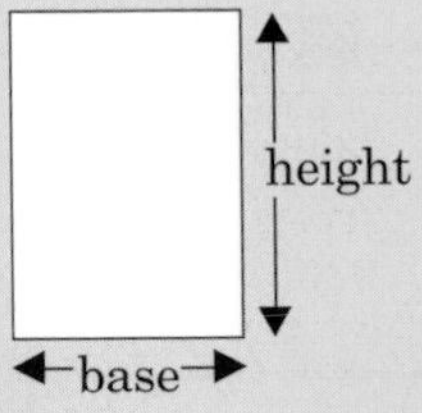

- You can make any side the base.
- Here, the rectangle is turned so that a longer side is the base:

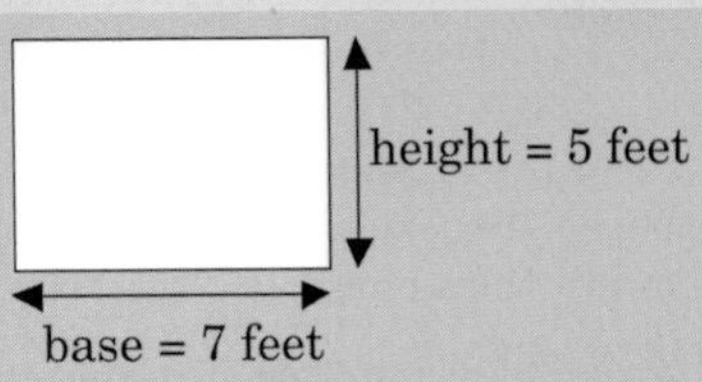

- Here, the same rectangle is turned so that a shorter side is the base:

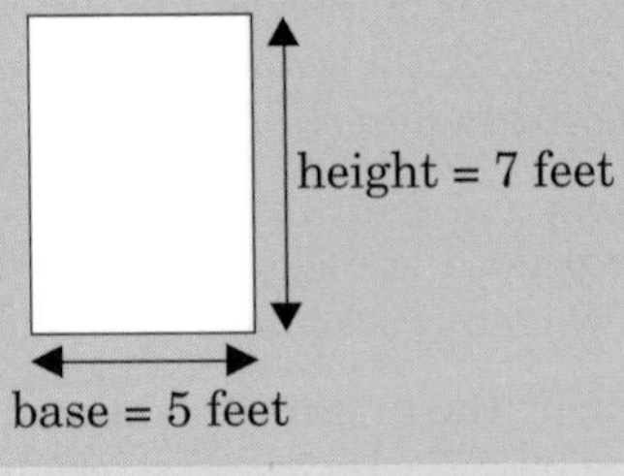

- Here's the equation for finding the area of a rectangle:
- Here's the multiplication:
- The answer is **square** units. The area of the figure is 35 square feet.

Area = base x height
A = b x h
A = 5 x 7

A = 35 square feet

- Here's the figure with the square feet visible:
- If you count them, you'll see that there are 35 squares inside the figure.

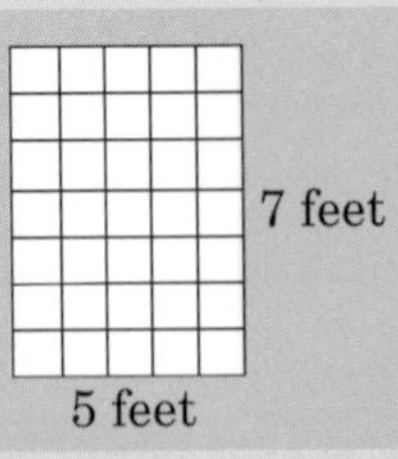

- Remember, the area of figures is always measured in **square** units.

Part 2 Figure out the area of each rectangle. Remember the unit name.

Area = base x height

a. 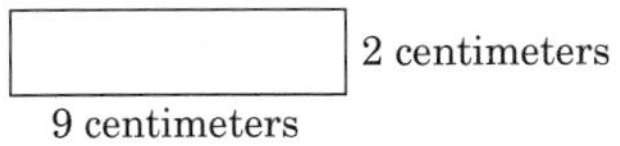

b. 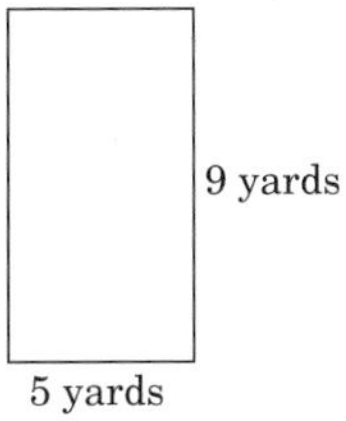

c. 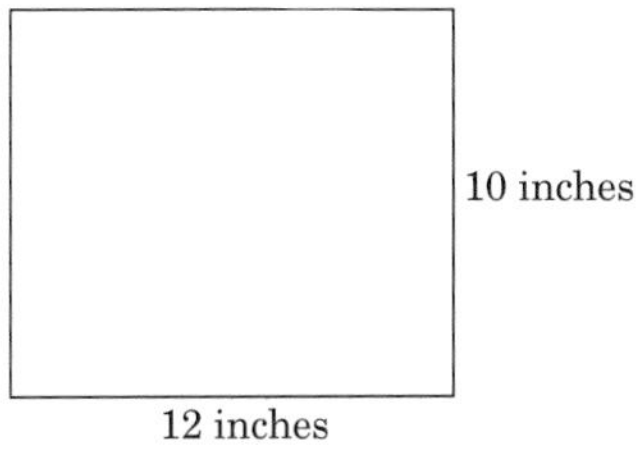

Part 3 Copy each mixed number. Write the fraction it equals.

a. $3\frac{7}{8}$ b. $1\frac{7}{9}$ c. $6\frac{2}{3}$ d. $5\frac{3}{7}$

Part 4 For each item, rewrite the equation. Figure out what the letter equals.

a. 18 x 9 = R b. 850 – 196 = D c. 117 + 685 = M

Part 5 For each item, make a number family with three names and two numbers. Then figure out the missing number.

a. 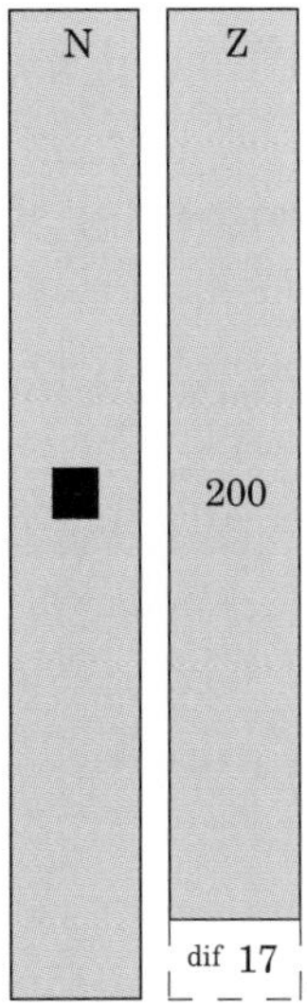

b. 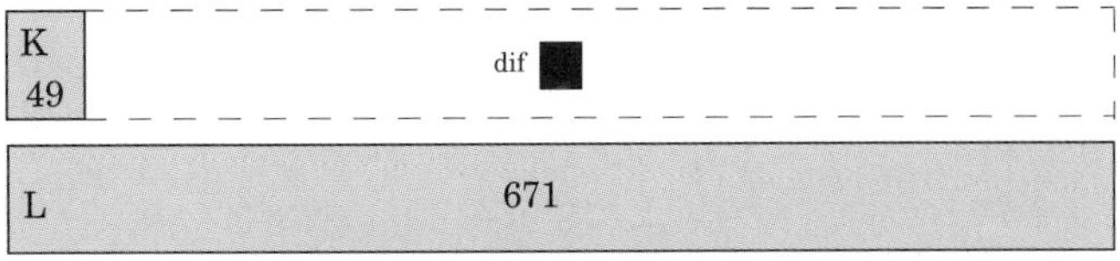

c. 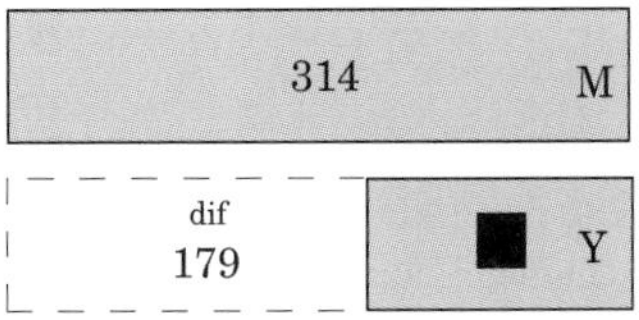

Part 6 Copy the table. Then figure out all the missing numbers.

			total
	120	■	365
	■	45	■
total	270	■	■

Part 7 Answer each question.

a. This table shows the number of horses and cows on a ranch.

	brown	not brown	total for both colors
horses	18	47	65
cows	115	63	178
total for both animals	133	110	243

b. This table shows the number of cars and trucks on two different lots.

	lot A	lot B	total for both lots
cars	132	125	257
trucks	16	64	80
total for both vehicles	148	189	337

a. What's the total number of cars for both lots?

b. How many vehicles are on lot B?

c. How many trucks are on lot B?

d. How many trucks are on lot A?

e. What's the total number of vehicles for both lots?

Independent Work

Part 8 **Copy and complete the ratio equation for each problem. Answer the question.**

a. There were 4 rabbits for every 7 carrots. If there were 98 carrots, how many rabbits were there? $\frac{\text{rabbits}}{\text{carrots}} \; \frac{4}{7} = \frac{■}{98}$

b. The ratio of nails to screws in a house was 11 to 2. If there were 120 screws in the house, how many nails were there? $\frac{\text{nails}}{\text{screws}} \; \frac{11}{2} = \frac{■}{120}$

c. A machine made 12 cards every 5 seconds. How many seconds would it take the machine to make 60 cards? $\frac{\text{cards}}{\text{seconds}} \; \frac{12}{5} = \frac{60}{■}$

Part 9 **For each problem, write the complete equation showing the fraction and the whole number it equals.**

a. The denominator of a fraction is 10. The fraction equals 6.

b. The denominator of a fraction is 6. The fraction equals 11.

c. $\frac{12}{1} = ■$ d. $30 = \frac{■}{5}$ e. $\frac{4}{4} = ■$ f. $\frac{■}{8} = 16$

Part 10 **Make a number family for each problem. Answer the question.**

a. A large building had 456 windows. 56 of them had tinted glass. How many did not have tinted glass?

b. There are 52 cards in a deck of cards. 13 of those cards are diamonds. How many cards are not diamonds?

c. Some of the cats in the animal hospital had fleas. There were 16 cats with fleas and 42 cats that did not have fleas. How many cats were in the hospital?

Part 11 **Copy the table and fill in the missing numbers.**

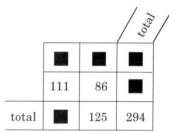

			total
	■	■	■
	111	86	■
total	■	125	294

Part 12 **Copy the table and complete all the rows.**

	Multiplication	Division	Fraction equation
a.	36 x ■ = 1620		
b.			$\frac{1196}{13} = ■$
c.		$28\overline{)448}$	

Part 13 **Answer each question.**

a. Each broom cost $4. A warehouse spent $68 for brooms. How many brooms did the warehouse buy?

b. There are 9 oranges in each bag. A truck contains 108 bags. How many oranges are in the truck?

Lesson 18

Part 1 **Answer each question.**

This table shows the number of penguins and seals observed on Black Beach and Surf Beach one day in December.

	penguins	seals	total for both animals
Black Beach	47	285	332
Surf Beach	108	95	203
total for both beaches	155	380	535

a. What was the total number of seals for both beaches?

b. How many penguins were observed on Surf Beach?

c. What was the total number of both animals on Black Beach?

d. On which beach were fewer seals observed?

e. 108 tells about the number of [redacted] on [redacted].

f. 95 tells about the number of [redacted] on [redacted].

g. 332 tells about the number of [redacted] on [redacted].

I see you have a table problem.
No, the table works fine. I'm having trouble with the math.
Town A
Town B
Total
15
17
52
99

Part 2

Read each item. If the equation is correct, copy it. If the equation is not correct, write the correct equation. Then answer the question.

a. On a train, there were 2 women for every 5 men. There were 48 women on the train. How many men were on the train?

$$\frac{\text{women}}{\text{men}} \quad \frac{2}{5} = \frac{48}{\blacksquare}$$

b. The ratio of cows to bulls on the farm was 3 to 1. If there were 15 bulls, how many cows were there?

$$\frac{\text{cows}}{\text{bulls}} \quad \frac{1}{3} = \frac{\blacksquare}{15}$$

c. The machine made 6 buttonholes every 17 seconds. How many buttonholes did the machine make in 85 seconds?

$$\frac{\text{buttonholes}}{\text{seconds}} \quad \frac{6}{17} = \frac{85}{\blacksquare}$$

d. The ratio of leaves on the ground to leaves in the tree was 2 to 7. If there were 308 leaves in the tree, how many leaves were on the ground?

$$\frac{\text{leaves on the ground}}{\text{leaves in the tree}} \quad \frac{2}{7} = \frac{\blacksquare}{308}$$

Part 3

For each item, make a number family with three names and two numbers. Figure out the missing number.

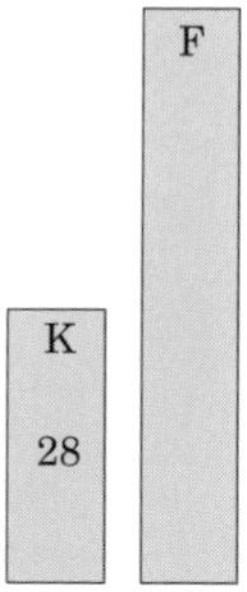

a. F is 31 more than K. How much is F?

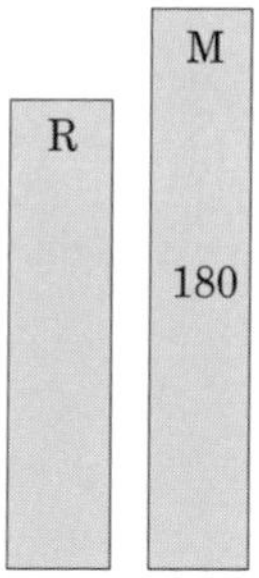

c. M is 16 more than R. How much is R?

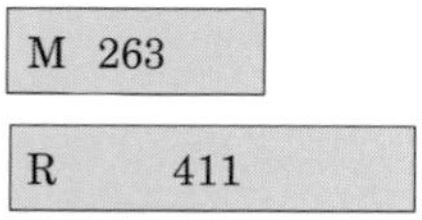

b. How much less is M than R?

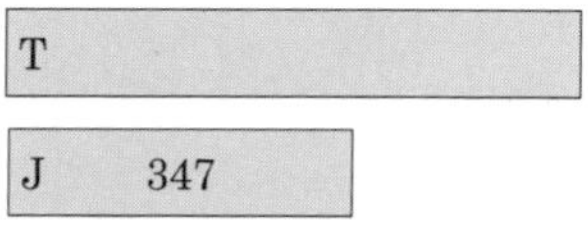

d. J is 134 less than T. How much is T?

Part 4 Copy the table. Then figure out all the missing numbers.

			total
	126	$\blacksquare$	$\blacksquare$
	$\blacksquare$	$\blacksquare$	154
total	$\blacksquare$	123	302

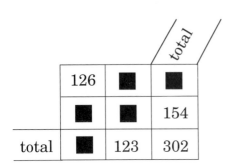

Part 5 Figure out the area of each rectangle.

a. 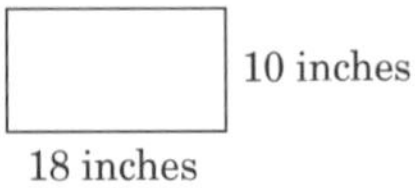

b. 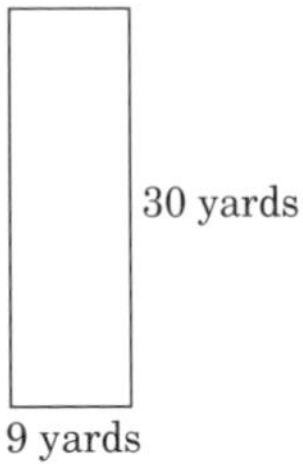

c. 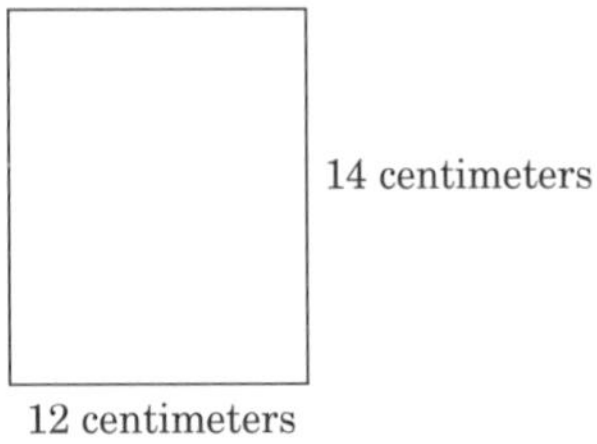

Part 6 Copy and complete each equation.

a. $1\frac{6}{7} = \blacksquare$ b. $2\frac{1}{6} = \blacksquare$ c. $2\frac{3}{5} = \blacksquare$ d. $3\frac{1}{4} = \blacksquare$ e. $4\frac{1}{3} = \blacksquare$

Independent Work

Part 7 For each item, write the complete equation showing the fraction and the whole number it equals.

a. $\frac{\blacksquare}{8} = 4$ b. $\frac{21}{3} = \blacksquare$ c. $\frac{\blacksquare}{3} = 12$ d. $5 = \frac{\blacksquare}{2}$ e. $\frac{15}{5} = \blacksquare$ f. $1 = \frac{\blacksquare}{5}$

Part 8 Work each item.

a. $\frac{3}{4}\left(\blacksquare\right) = \frac{9}{20}$

b. $37 - \blacksquare = 0$

c. $\frac{16}{5} + \frac{3}{5} = \blacksquare$

d. $2 \times \blacksquare = 2$

e. $\frac{14}{5} - 1 = \blacksquare$

f. $\frac{12}{3} \times \frac{1}{3} = \blacksquare$

g. $\frac{7}{3} = \frac{\blacksquare}{36}$

h. $\frac{6}{7} = \frac{54}{\blacksquare}$

Part 9 Find the perimeter of each figure.

a. 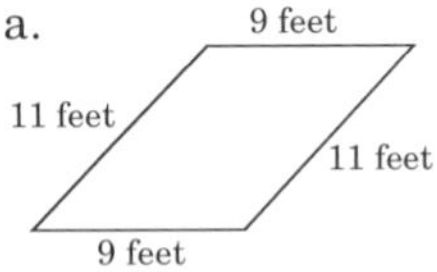

b. 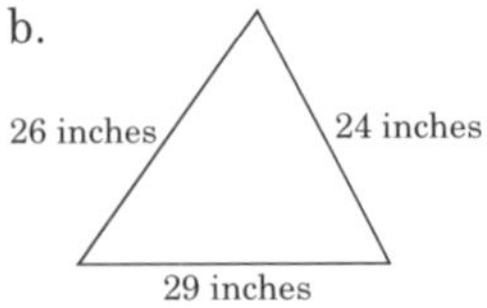

c. 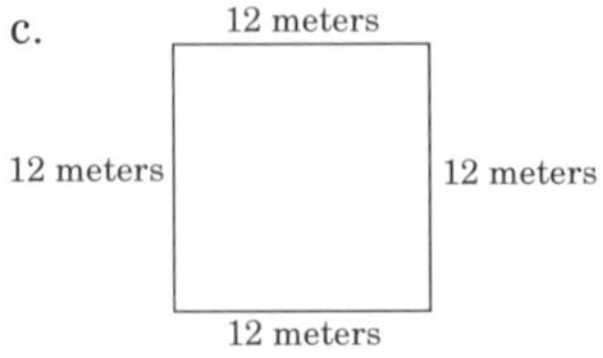

Part 10

Copy and complete each item.

a. 5 (■) = 960

b. 34 (■) = 884

c. 7 (■) = 1309

d. 8 (■) = 192

Part 11

Answer each question.

a. Each marble weighs 8 grams. There are 35 marbles. What is their weight?

b. A large parcel of land is divided into 9 smaller parcels. Each smaller parcel is the same size. If the larger parcel is 630 acres, how large is each smaller parcel?

c. A machine makes 34 strips each minute. How many strips will the machine make in 15 minutes?

d. A pie is cut into 5 pieces that are the same size. Each piece weighs 6 ounces. How much does the entire pie weigh?

Part 12

For each number line, write the complete equation. Show the mixed number and the fraction it equals.

a. 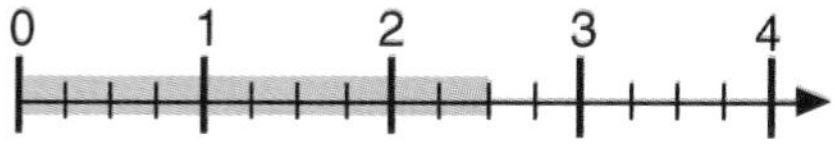

b. 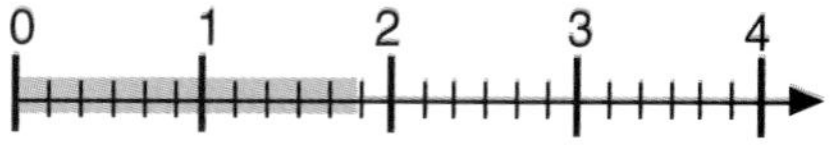

c. 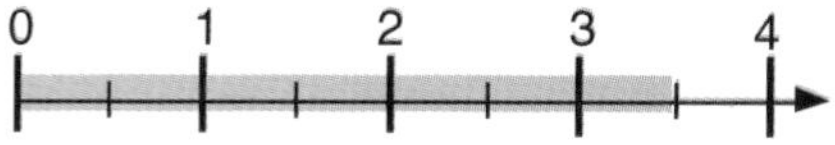

Part J

a. $\frac{\text{women}}{\text{men}}\ \frac{2}{5}\left(\frac{24}{24}\right) = \frac{48}{120}$ [120] [120 men]

b. $\frac{\text{cows}}{\text{bulls}}\ \frac{3}{1}\left(\frac{15}{15}\right) = \frac{45}{15}$ [45] [45 cows]

c. $\frac{\text{buttonholes}}{\text{seconds}}\ \frac{6}{17}\left(\frac{5}{5}\right) = \frac{30}{85}$ [30] [30 buttonholes]

d. $\frac{\text{leaves on the ground}}{\text{leaves in the tree}}\ \frac{2}{7}\left(\frac{44}{44}\right) = \frac{88}{308}$ [88] [88 leaves]

Lesson 19

Part 1 Answer each question.

This table shows the number of cows and bulls on Blue Ranch and Russell Ranch.

a. How many cows are on Russell Ranch?

b. On which ranch are there more bulls?

c. What's the total number of bulls for both ranches?

d. 376 tells about the number of [redacted] on [redacted].

e. 903 tells about the number of [redacted] on [redacted].

f. 189 tells about the number of [redacted] on [redacted].

	cows	bulls	total for both animals
Blue Ranch	618	189	807
Russell Ranch	285	376	661
total for both ranches	903	565	1468

Part 2 Copy and complete each pair of equivalent fractions.

a. $\frac{2}{5} = \frac{\text{[redacted]}}{940}$

b. $\frac{3}{14} = \frac{\text{[redacted]}}{1022}$

c. $\frac{8}{7} = \frac{728}{\text{[redacted]}}$

d. $\frac{15}{4} = \frac{465}{\text{[redacted]}}$

Part 3 For each item, make a number family with three names and two numbers. Figure out the missing number.

a.

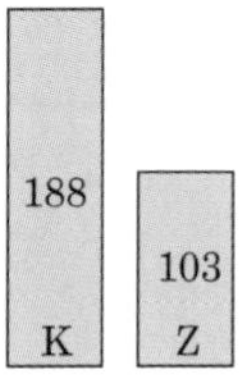

How much less is Z than K?

b.

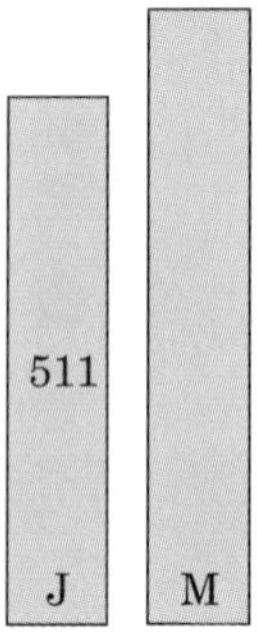

J is 68 less than M. How much is M?

c.

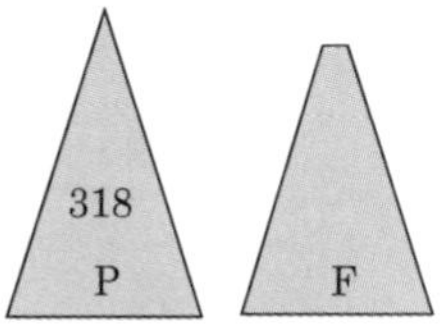

P is 17 more than F. How much is F?

d.

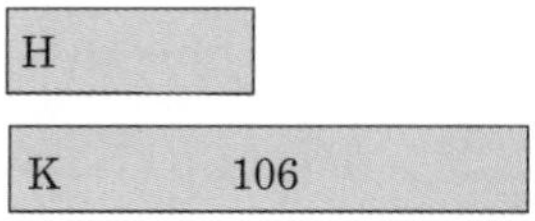

H is 56 less than K. How much is H?

Part 4 For each item, write the complete ratio equation with names. Then figure out the missing number and answer the question.

a. In the forest, the ratio of tall trees to short trees was 2 to 7. There were 120 tall trees in the forest. How many short trees were in the forest?

b. A tire travels 8 yards every 3 turns. How many yards does the tire travel in 48 turns?

c. In Newton School, there are 4 girls to every 9 students. There are 189 students in Newton School. How many girls go to Newton School?

Part 5 Find the perimeter and the area of each rectangle.

a.

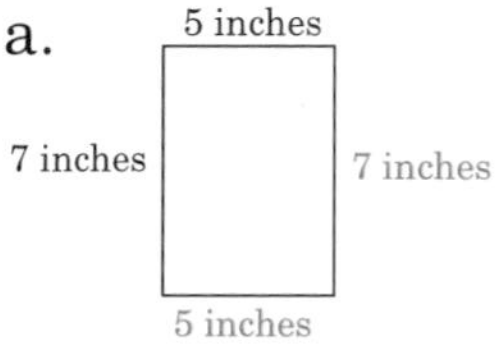

b.

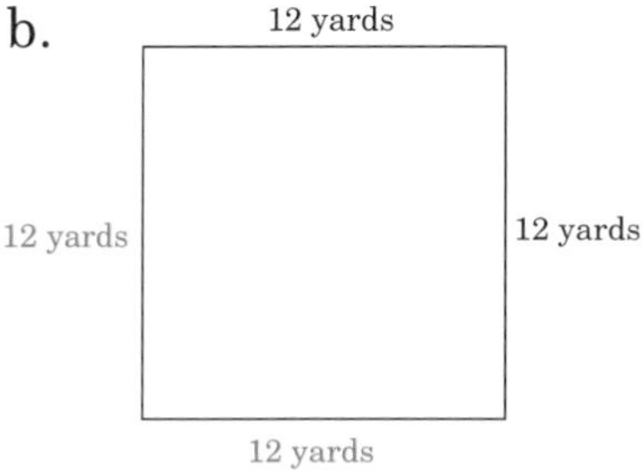

c.

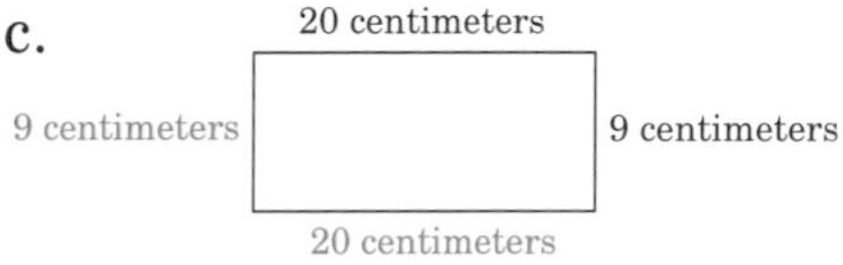

Independent Work

Part 6 **Work each item.**

a. $\frac{437}{19} =$ [redacted]   b. $\frac{6}{4} = \frac{[redacted]}{80}$ c. $\frac{8}{4} + 1 =$ [redacted]   d. $\frac{3}{7} \times \frac{3}{7} =$ [redacted]   e. $\frac{2}{7} = \frac{[redacted]}{98}$

f. $\frac{495}{55} =$ [redacted]   g. $\frac{8}{5} - \frac{5}{5} =$ [redacted]   h. $\frac{270}{45} =$ [redacted]   i. $\frac{[redacted]}{3} = 36$

Part 7 **Figure out what to multiply the first fraction by. If the fractions shown are equivalent, write the simple equation below.**

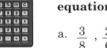

a. $\frac{3}{8}, \frac{24}{64}$ b. $\frac{5}{4}, \frac{100}{80}$

c. $\frac{3}{5}, \frac{60}{80}$ d. $\frac{7}{5}, \frac{350}{500}$

Part 8 **For each item, make a number family. Answer the question.**

a. A large company had 17 trucks that needed repair and 236 trucks that did not need repair. How many trucks did the company have?

b. A bike shop had 23 tires that needed repair. If the shop had a total of 98 tires, how many tires did not need repairs?

c. There were 123 young children at a picnic. At the end of the day, 107 were dirty. How many were still clean?

d. When Jan's family went on their vacation, they had 15 cloudy days and 11 days that were not cloudy. How many days was the family on vacation?

Part 9 **Write equations that show each mixed number and the fraction it equals.**

a. $3\frac{4}{10}$ b. $7\frac{1}{2}$ c. $1\frac{24}{30}$

d. $10\frac{7}{9}$ e. $8\frac{1}{6}$

Part 10 **Rewrite each equation. Write what the letter equals.**

a. 560 + 13 = J

b. 816 – 539 = T

c. 56 + 139 = B

d. 32 x 11 = P

Part 11 **Copy and complete the table.**

			total
	270	[redacted]	368
	[redacted]	[redacted]	268
total	336	[redacted]	636

Part J

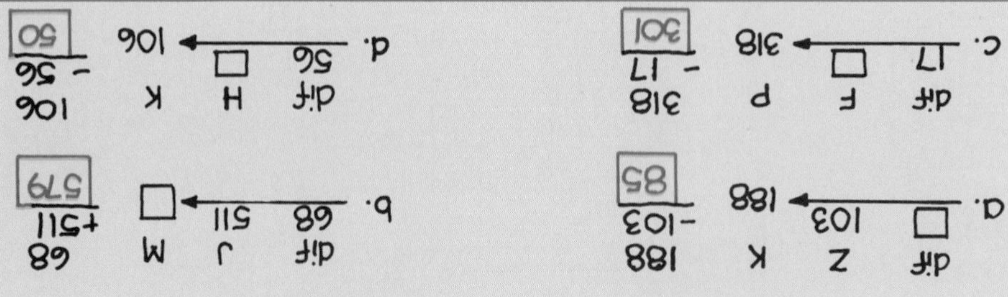

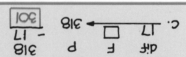

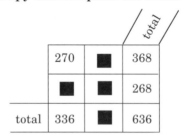

Lesson 20

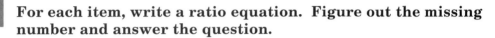

Part 1 **For each item, write a ratio equation. Figure out the missing number and answer the question.**

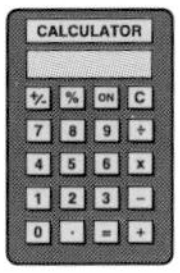

a. In a library, there were 6 red books for every 7 blue books. The library had 497 blue books. How many red books did the library have?

b. In a marsh, there were 5 ducks for every 4 geese. 72 geese lived in the marsh. How many ducks lived in the marsh?

c. A rod is a unit for measuring length. The ratio of yards to rods is 11 to 2. If something is 44 rods long, how many yards long is that object?

d. The ratio of rabbits to carrots is 8 to 3. If there are 48 rabbits, how many carrots are there?

Part 2 **If the equation is correct, copy it. If the equation is not correct, write the correct equation. Complete each equation and answer the question.**

a. In a marsh, the ratio of fish to birds is 3 to 4. There are 600 fish in the swamp. How many birds are in the swamp?

$$\frac{\text{fish}}{\text{birds}} \; \frac{3}{4} = \frac{600}{\blacksquare}$$

b. If 2 bricks weigh 9 pounds, how many bricks weigh 252 pounds?

$$\frac{\text{bricks}}{\text{pounds}} \; \frac{2}{9} = \frac{\blacksquare}{252}$$

c. If it takes 3 workers to move 2 tons, how many workers are needed to move 74 tons?

$$\frac{\text{workers}}{\text{tons}} \; \frac{3}{2} = \frac{74}{\blacksquare}$$

d. In a lake, the ratio of bass to trout is 4 to 7. There are 112 bass. How many trout are in the lake?

$$\frac{\text{bass}}{\text{trout}} \; \frac{7}{4} = \frac{112}{\blacksquare}$$

Part J

a. $\frac{\text{red books}}{\text{blue books}}\ \frac{6}{7}\left(\frac{71}{71}\right) = \frac{426}{497}$ — 426 red books

b. $\frac{\text{ducks}}{\text{geese}}\ \frac{5}{4}\left(\frac{18}{18}\right) = \frac{90}{72}$ — 90 ducks

c. $\frac{\text{yards}}{\text{rods}}\ \frac{11}{2}\left(\frac{22}{22}\right) = \frac{242}{44}$ — 242 yards

d. $\frac{\text{rabbits}}{\text{carrots}}\ \frac{8}{3}\left(\frac{6}{6}\right) = \frac{48}{18}$ — 18 carrots

Part K

a. $\frac{\text{fish}}{\text{birds}}\ \frac{3}{4}\left(\frac{200}{200}\right) = \frac{600}{800}$ — 800 birds

b. $\frac{\text{bricks}}{\text{pounds}}\ \frac{2}{9}\left(\frac{28}{28}\right) = \frac{56}{252}$ — 56 bricks

c. $\frac{\text{workers}}{\text{tons}}\ \frac{3}{2}\left(\frac{37}{37}\right) = \frac{111}{74}$ — 111 workers

d. $\frac{\text{bass}}{\text{trout}}\ \frac{4}{7}\left(\frac{28}{28}\right) = \frac{112}{196}$ — 196 trout

Test 2

Part 1 **Copy and complete each equation.**

a. $\frac{18}{2} = ■$ b. $\frac{■}{10} = 10$ c. $■ = \frac{9}{3}$ d. $12 = \frac{■}{6}$

Part 2 **Rewrite each equation so the letter is first. Then write what the letter equals.**

a. 56 x 4 = T b. 96 – 41 = P c. 860 + 800 = J

Part 3 **Copy and complete each equation.**

a. $\frac{5}{8} = \frac{35}{■}$ b. $\frac{7}{5} = \frac{56}{■}$ c. $\frac{8}{9} = \frac{■}{90}$

Part 4 For each problem, make the number family with names. Write the answer as a number and a unit name.

a. In a lunchroom, there were 16 tables that were clean and 24 tables that were dirty. How many tables were in the lunchroom?

b. Mr. James went on a trip. In the morning, he traveled 234 miles. In the afternoon, he traveled farther. If the entire trip was 660 miles, how far did Mr. James travel in the afternoon?

c. The horses ate 165 pounds of grain in January. In February, they ate 189 pounds of grain. How much grain did they eat during the two-month period?

Part 5 Copy and complete the table.

	Multiplication	Division	Fraction equation
a.			$\frac{74}{2} =$ ■
b.	7 x ■ = 231		
c.	5 x ■ = 95		
d.			$\frac{21}{3} =$ ■

Part 6 Find the perimeter of each figure. Write the answer as a number and unit name.

a.

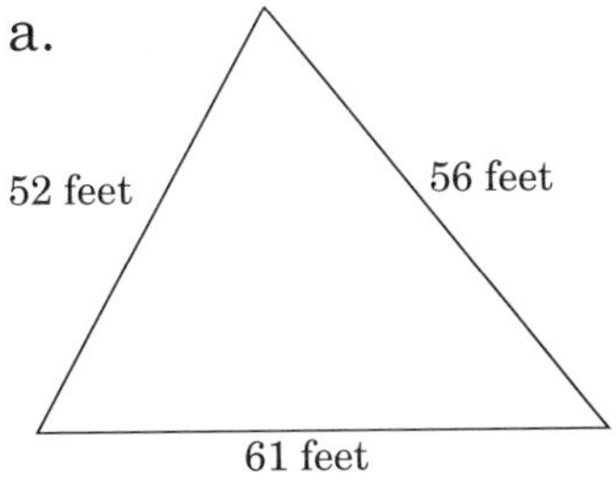

b.

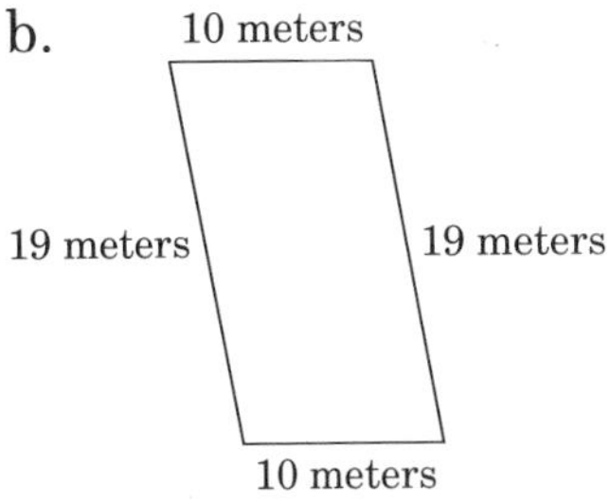

Part 7

For each item, write the equation to show the mixed number and the fraction it equals.

a. $3\frac{4}{9} =$ ■ b. $3\frac{7}{10} =$ ■ c. $4\frac{1}{8} =$ ■ d. $4\frac{4}{5} =$ ■

Part 8

For each item, make a number family that shows the difference. Then figure out answer.

a.

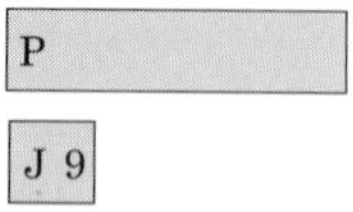

P is 36 more than J.
How much is P?

b.

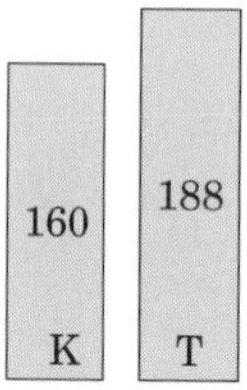

How much less is K than T?

c.

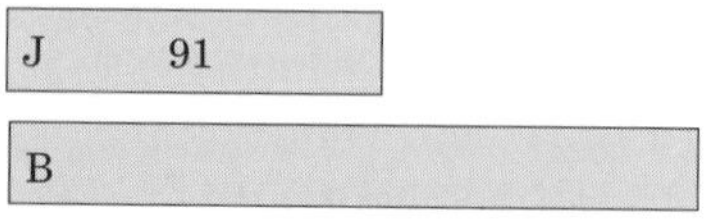

J is 58 less than B.
How much is B?

Part 9

Copy the table. Then figure out all the missing numbers.

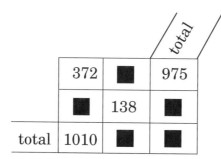

			total
	372	■	975
	■	138	■
total	1010	■	■

Lesson 21

Part 1 Copy and complete each equation.

a. $\frac{7}{15} = \frac{847}{\text{[redacted]}}$

b. $\frac{10}{3} = \frac{\text{[redacted]}}{669}$

c. $\frac{7}{8} = \frac{\text{[redacted]}}{640}$

d. $\frac{5}{17} = \frac{625}{\text{[redacted]}}$

Part 2 For each item, make a number family with names and numbers. Then figure out the missing number.

a.

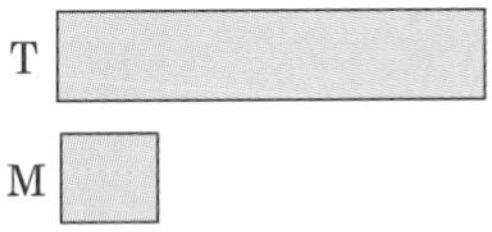

Bar T is 143 meters long. Bar M is 112 meters shorter than bar T. How long is bar M?

b.

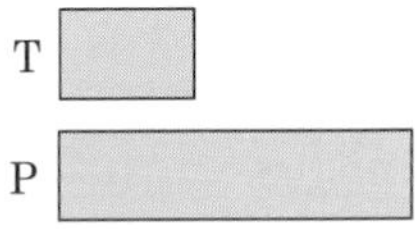

Bar P is 118 inches long. Bar T is 45 inches long. How much longer is bar P than bar T?

c.

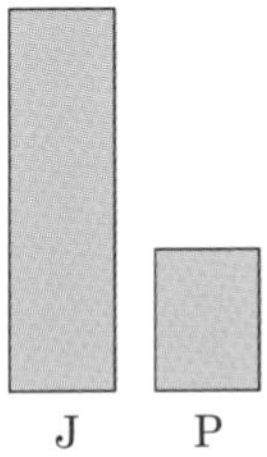

Bar J is 77 feet longer than bar P. Bar P is 45 feet long. How long is bar J?

d.

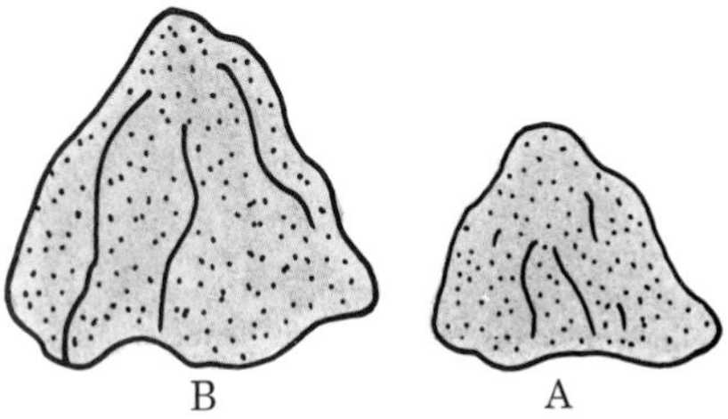

Pile A weighs 456 pounds less than pile B. Pile A weighs 545 pounds. How much does pile B weigh?

Part 3

- Some lines **intersect.** Those are lines that touch each other or cross each other.
- In each of these pictures, lines intersect.

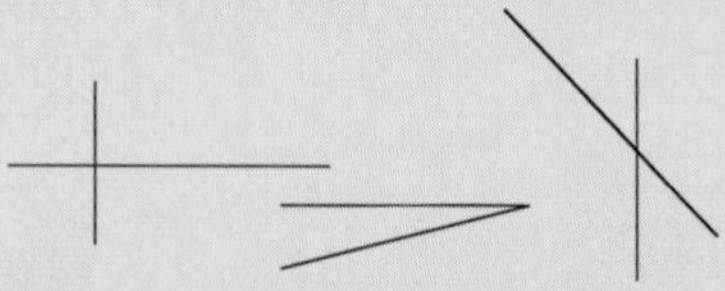

- Some lines do not intersect, but they would intersect if the lines were extended.
- In each of these pictures, the lines would intersect if they were extended.

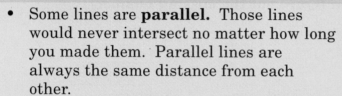

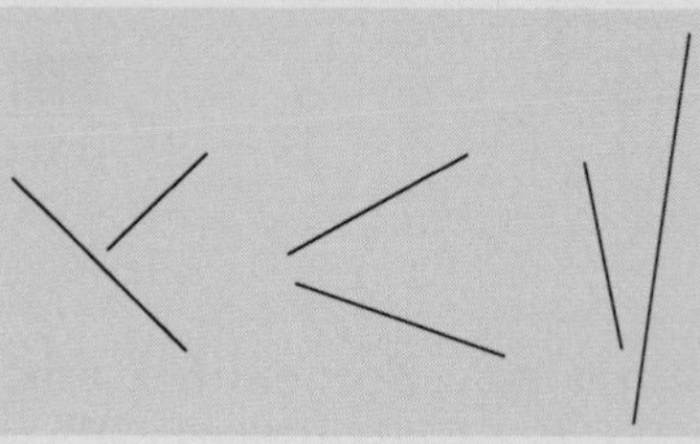

- Some lines are **parallel.** Those lines would never intersect no matter how long you made them. Parallel lines are always the same distance from each other.
- In each of these pictures, the lines are parallel. So they do not intersect.

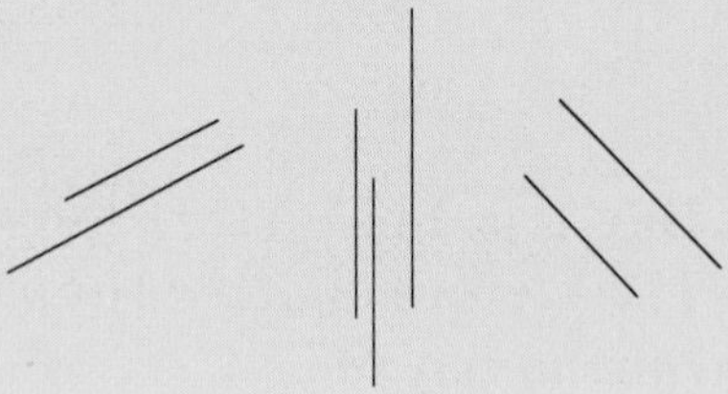

Part 4

Write whether each set of lines is parallel or not parallel.

a. 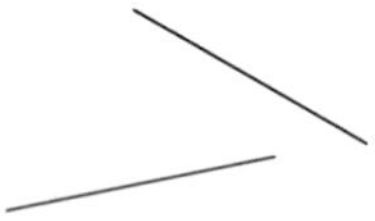

b. 

c. 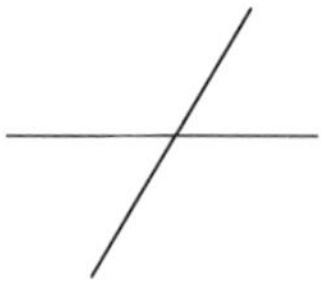

d. 

e. 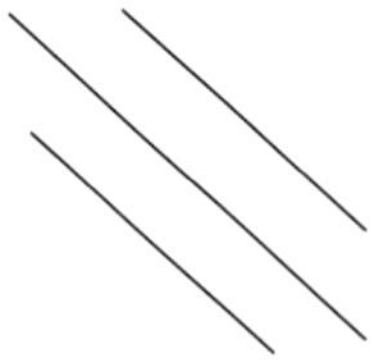

f. 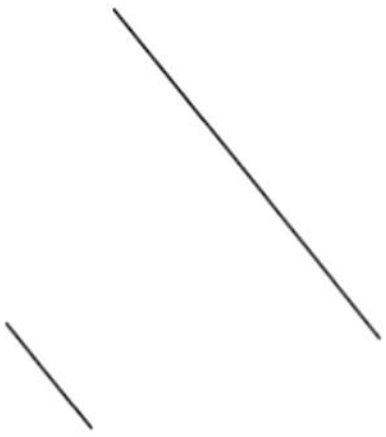

Part 5 **Figure out the perimeter and area of each rectangle. Use abbreviations for the unit names.**

a.

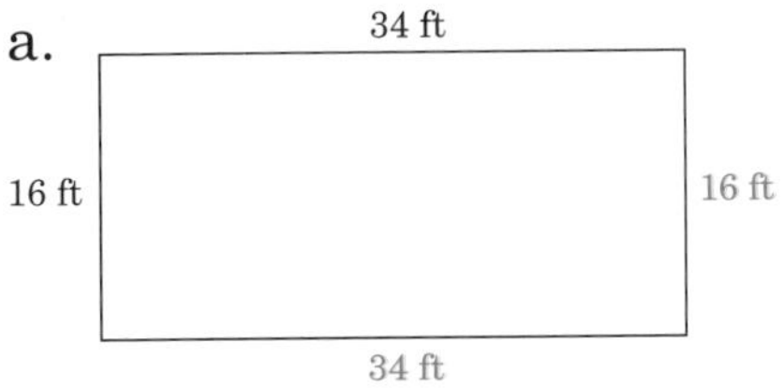

b.

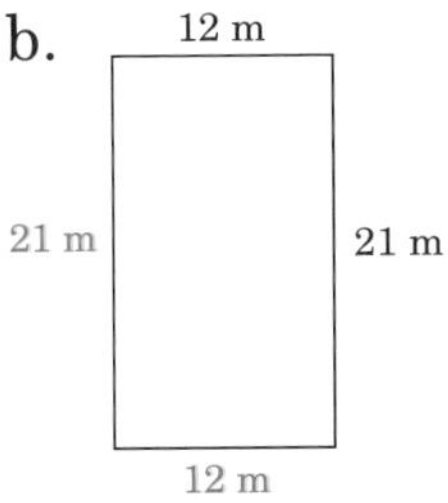

c.

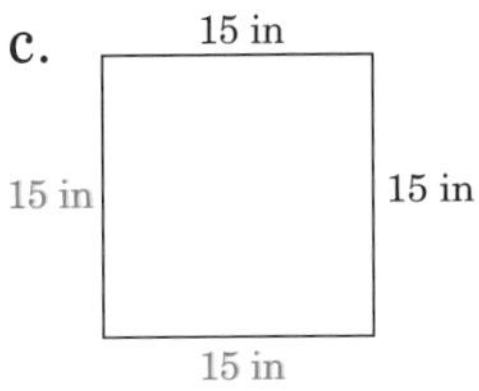

Part 6 **Copy the table. Then figure out all the missing numbers.**

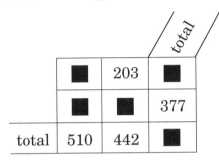

			total
	[redacted]	203	[redacted]
	[redacted]	[redacted]	377
total	510	442	[redacted]

Part 7 **Answer each question.**

This table shows the number of ripe grapes and unripe grapes in two gardens.

	ripe	unripe	total grapes
Brandon's garden	950	104	1054
Eden's garden	708	450	1158
total for both gardens	1658	554	2212

a. Which garden has more grapes that are ripe?

b. What is the total number of grapes for both gardens?

c. Which is greater, the total for ripe grapes or the total for grapes that are unripe?

d. 1158 tells about the number of [redacted] in [redacted].

e. 104 tells about the number of [redacted] in [redacted].

Independent Work

Part 8

For each problem, write the names and a complete equation. Box your answer with a unit name.

a. In a bakery, the ratio of cookies to muffins is 4 to 3. There are 92 cookies in the bakery. How many muffins are there?

b. Joe writes 9 reports every 5 days. He wrote for 20 days in April. How many reports did Joe write in April?

c. It takes 2 loads for a machine to wash 19 towels. How many towels can the machine wash in 38 loads?

Part 9

Figure out what to multiply the first fraction by to get the second fraction. If the fractions shown are equivalent, write the simple equation below.

a. $\frac{2}{7}$, $\frac{24}{28}$ b. $\frac{3}{4}$, $\frac{48}{64}$ c. $\frac{9}{5}$, $\frac{36}{20}$

Part 10

Copy each problem. Write the answer with a remainder.

a. $7\overline{)30}$ b. $5\overline{)48}$

c. $2\overline{)43}$ d. $4\overline{)55}$

Part 11

Write equations that show each mixed number and the fraction it equals.

a. $5\frac{3}{7}$ b. $1\frac{6}{9}$

c. $8\frac{1}{2}$ d. $13\frac{2}{5}$

Part 12

For each item, make a number family. Answer the question the problem asks.

a. There are 189 children in a school. 74 of them are boys. How many are girls?

b. A truck has a load of gravel that weighs 69,000 pounds. 45,200 pounds of this gravel was in larger chunks of gravel. How many pounds of smaller chunks were there?

c. A company owned vehicles that needed repair and vehicles that did not need repair. 82 vehicles needed repair. 356 did not. How many vehicles did the company own?

d. Painters used red paint and other colors to paint a building. They used 69 gallons of paint in all. 19 gallons were red paint. How many gallons of other paint did they use?

Part 13

Work each item.

a. $77 - \blacksquare = 0$

b. $1 - \frac{4}{5} = \blacksquare$

c. $\frac{8}{3} \times \frac{5}{3} = \blacksquare$

d. $\frac{16}{4} \times \frac{6}{4} = \blacksquare$

e. $20 \times \blacksquare = 20$

f. $47 \times 21 = \blacksquare$

g. $\frac{8}{3} = \frac{\blacksquare}{18}$

h. $\frac{5}{9}\left(\blacksquare\right) = \frac{35}{45}$

i. $\blacksquare + 1 = 1$

j. $603 - 57 = \blacksquare$

k. $\frac{36}{9} = \blacksquare$

l. $10 = \frac{\blacksquare}{5}$

m. $\frac{17}{17} - 1 = \blacksquare$

n. $\frac{6}{5} = \frac{30}{\blacksquare}$

o. $\frac{\blacksquare}{1} = 26$

Part 14 Copy the table and complete all the rows.

	Multiplication	Fraction equation	Division
a.	7(■) = 287	$\frac{■}{■}$ = ■	$\overline{)}$
b.		$\frac{965}{5}$ = ■	$\overline{)}$
c.		$\frac{■}{■}$ = ■	$20\overline{)480}$
d.	16(■) = 64	$\frac{■}{■}$ = ■	$\overline{)}$

Part J

b. dif T P: □ 45 → 118; 118 − 45 = **73 inches**

c. dif P J: 77 45 → □; 77 + 45 = **122 feet**

d. dif A B: 456 545 → □; 456 + 545 = **1001 pounds**

Part K

b. 21 + 21 + 12 + 12: **P = 66 m**; A = b × h, A = 12 × 21, **A = 252 sq m**

c. 15 + 15 + 15 + 15: **P = 60 in**; A = b × h, A = 15 × 15, **A = 225 sq in**

If you buy now, I'll give you an unbelievable bonus. Instead of getting just one video game for $12.00, I'll give you $\frac{5}{5}$ video games for the same low price.
Wow, that's great!
!

Lesson 22

Part 1 **Write whether each pair of lines is parallel or not parallel. Then write intersect or do not intersect for lines that are not parallel.**

a. 

b. 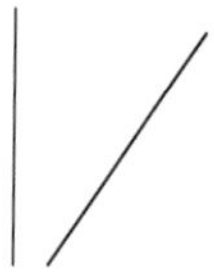

c. 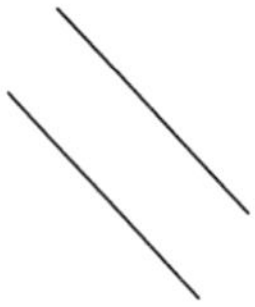

d. 

Part 2 **Work each item. Show your answer with a number and a unit name.**

a.

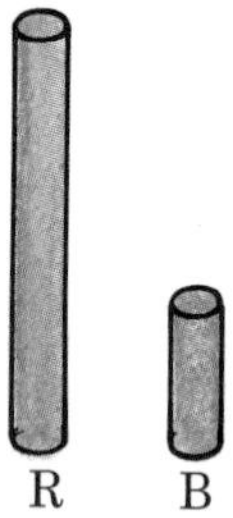

Pole B is 171 inches shorter than pole R. Pole B is 98 inches long. What is the length of pole R?

b.

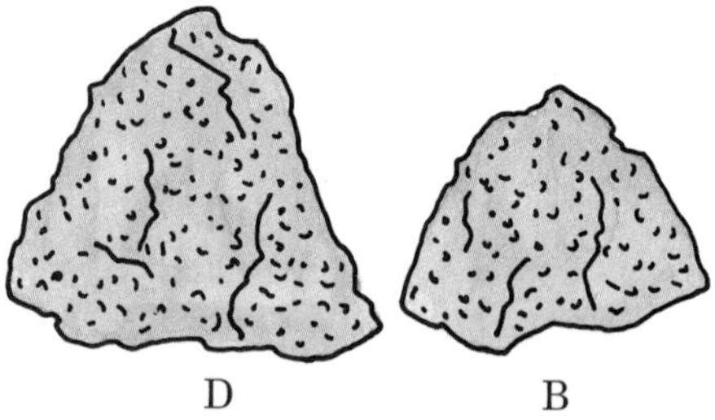

Pile B weighs 345 ounces. Pile D weighs 459 ounces. How much lighter is pile B than pile D?

c.

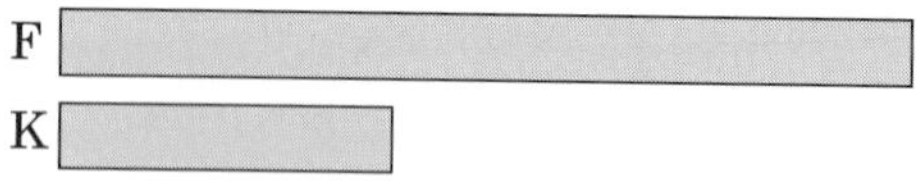

Bar F is 37 meters longer than bar K. Bar F is 199 meters long. How long is bar K?

d.

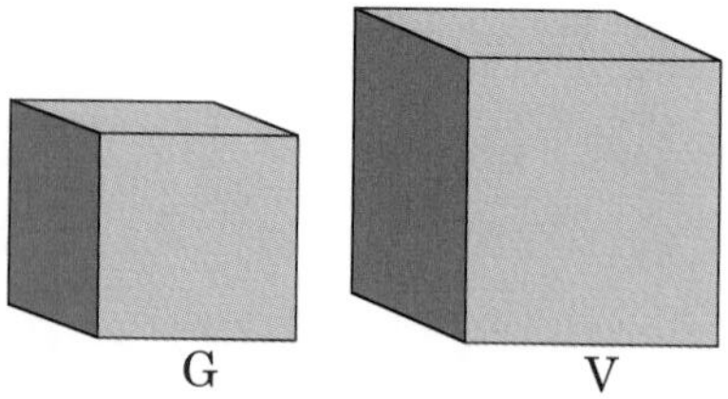

Box V is 62 pounds heavier than box G. Box G weighs 102 pounds. How much does box V weigh?

Part 3

- On some 4-sided figures, both pairs of opposite sides are parallel.
- Here are some of those figures:

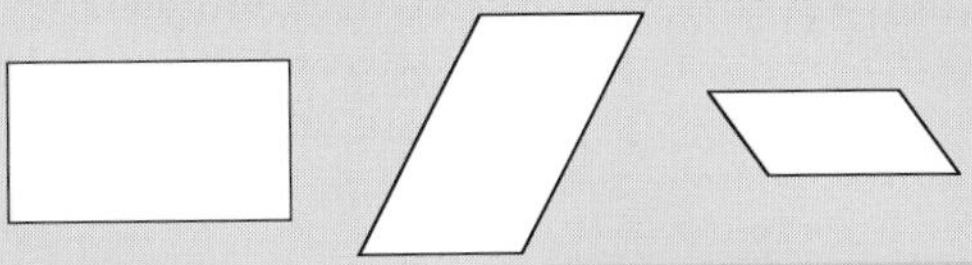

- On this figure, the left side is parallel

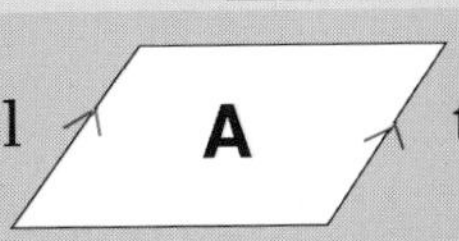

to the right side.

- And the top side is parallel

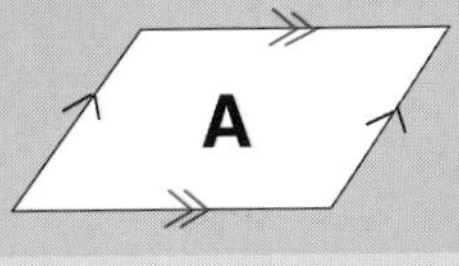

to the bottom side.

- Here's the rule:

> When both pairs of opposite sides are parallel, the two sides in each pair are the same length.

- The left side is the same length as the opposite side.
- The bottom side is the same length as the opposite side.

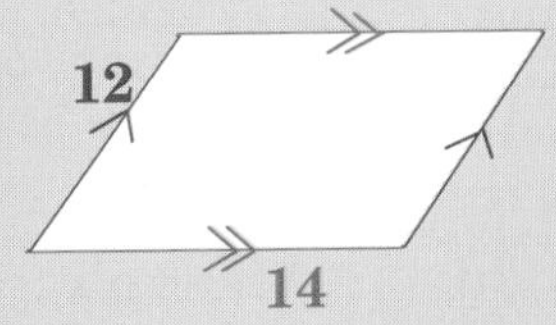

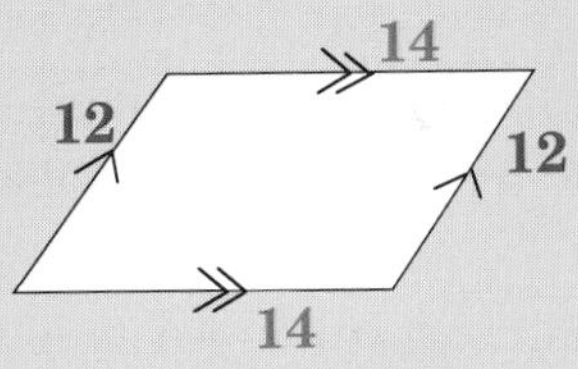

Part 4

Trace figures that have two pairs of parallel sides. Find the perimeter of those figures.

a.

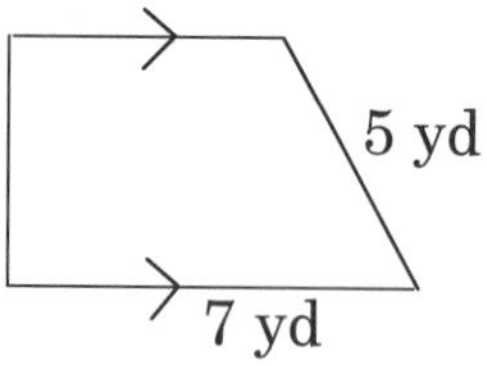

b.

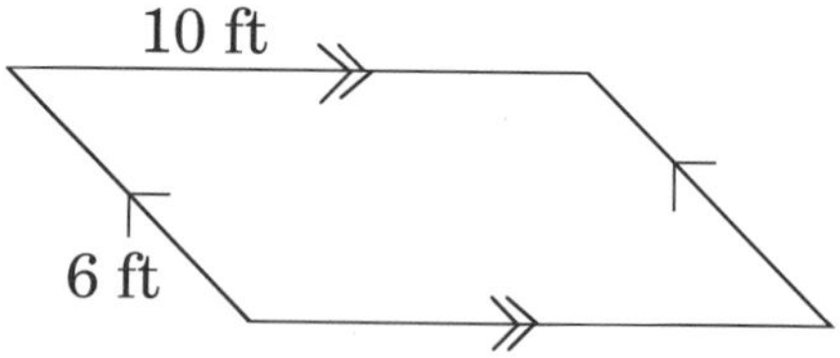

c.

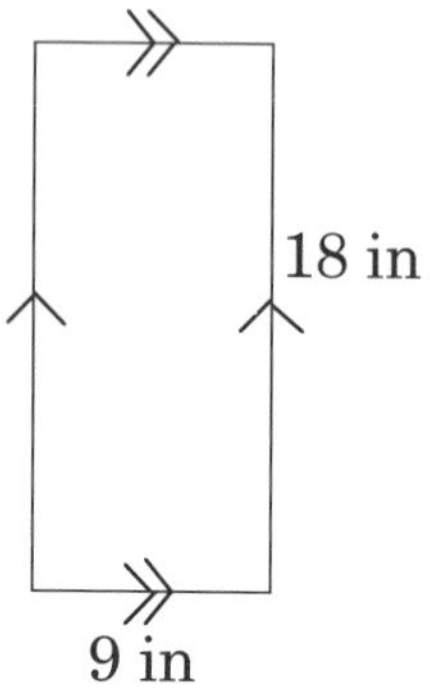

d.

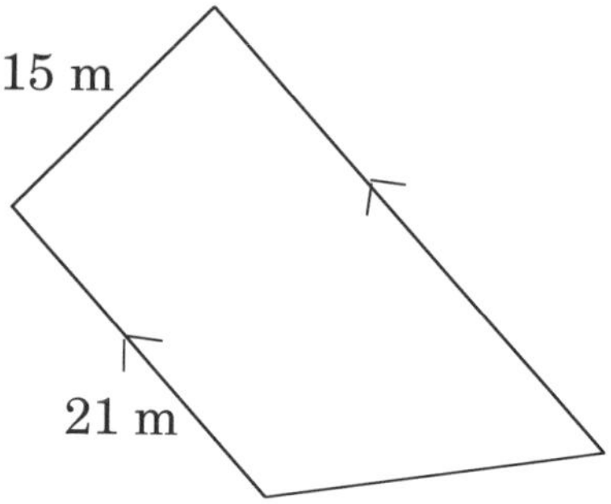

Part 5

Copy and complete the table. Then answer the questions.

This table is supposed to show the number of inches of snow that fell in Butte City and Ski Town in 1986 and 1987.

	1986	1987	total for both years
Butte City	48	[redacted]	154
Ski Town	78	[redacted]	[redacted]
total for both towns	[redacted]	[redacted]	319

a. In which year did more snow fall in Ski Town?

b. How many inches of snow fell in Butte City in 1986?

c. Which city had a higher total snowfall for both years?

d. What was the total number of inches of snow for both cities during both years?

e. 106 tells about the number of inches of snow that fell in [redacted] during [redacted].

Part 6 For each item, make a number family with three names and a difference number.

Sample sentences

1. Building M is 121 feet shorter than building C.
2. The pile of salt weighed 16 ounces more than the pile of baking powder.

a. James was 23 centimeters taller than Dan.

b. There were 89 more red trucks than yellow trucks.

c. Jan was 17 months older than her sister Ann.

d. The dog was 31 centimeters shorter than the snake.

e. The train went 48 miles per hour slower than the car.

Part 7 Copy the table and complete it.

	Mixed number	Fraction
a.	$4\frac{1}{2}$	
b.	$2\frac{4}{9}$	
c.	$3\frac{5}{8}$	
d.	$4\frac{2}{8}$	

Independent Work

Part 8 For each problem, write the names and a complete equation. Then box your answer with a unit name.

a. On a bush, the ratio of leaves to branches is 9 to 4. There are 144 leaves on the bush. How many branches are on the bush?

b. Every 4 bugs ate 6 seeds. A group of bugs ate 360 seeds. How many bugs were there?

c. A company uses 7 spools of cable every 2 miles. If the company lays cable from Portertown to Glenville, the company uses 56 spools of cable. How far apart are Portertown and Glenville?

Part 9 For each item, write the fraction and the whole number it equals.

a. This fraction equals 5 and the denominator is 12.

b. This fraction has a denominator of 1 and it equals 156.

c. This fraction has a denominator of 2 and it equals 156.

d. This fraction has a denominator of 4 and it equals 156.

Part 10 Work each item.

a. $\frac{■}{10} = 10$

b. $\frac{18}{3} + 1 = ■$

c. $\frac{45}{3} = ■$

d. $■ - 0 = 36$

e. $\frac{7}{8} = \frac{■}{64}$

f. $15 \times ■ = 0$

g. $\frac{2}{7} \times \frac{14}{1} = ■$

h. $■ + 64 = 65$

i. $1 - \frac{3}{10} = ■$

j. $\frac{17}{5} + \frac{7}{5} = ■$

k. $54 \times 12 = ■$

l. $\frac{9}{2}\left(■\right) = \frac{27}{24}$

m. $83 + 106 + 26 = ■$

n. $6 = \frac{■}{3}$

o. $\frac{1}{4} = \frac{■}{32}$

p. $■ = \frac{64}{2}$

q. $64 \times 173 = ■$

Part 11 Rewrite each equation so it begins with the letter. Below, write what the letter equals.

a. 152 – 17 = R

b. 28 + 29 = J

c. 200 – 186 = P

Part 12 Copy and work each problem.

a. $9\overline{)738}$

b. $6\overline{)462}$

c. $5\overline{)230}$

d. $3\overline{)183}$

Part 13 Find the perimeter of each figure.

a.

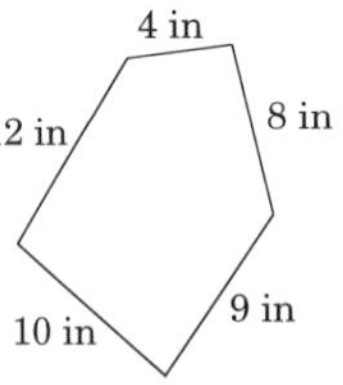

b.

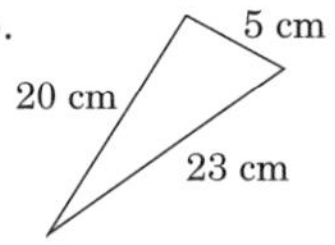

Part J

a. dif B R: 171, 98 → □; 171 + 98 = 269 in

b. dif B D: □, 345 → 459; 459 − 345 = 114 oz

c. dif K F: 37, □ → 199; 199 − 37 = 162 m

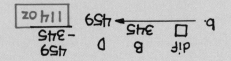

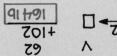

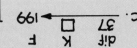

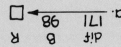

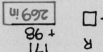

d. dif G V: 62, 102 → □; 62 + 102 = 164 lb

Lesson 23

Part 1 **Write whether each set of lines is parallel or not parallel. Then write intersect or do not intersect for lines that are not parallel.**

a.

b.

c.
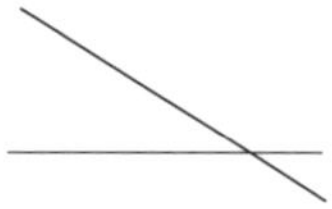

d.
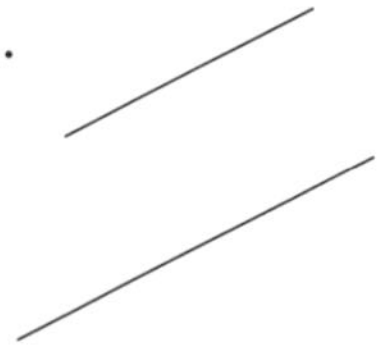

Part 2 **Trace figures that have two pairs of parallel sides. Find the perimeter of those figures.**

a.
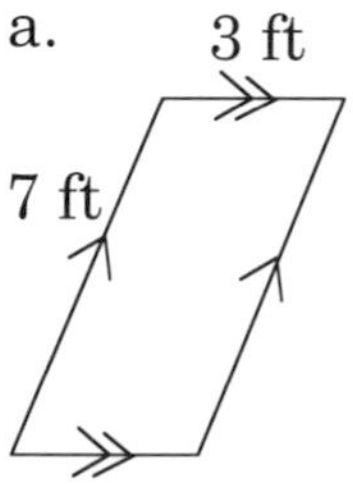

b.
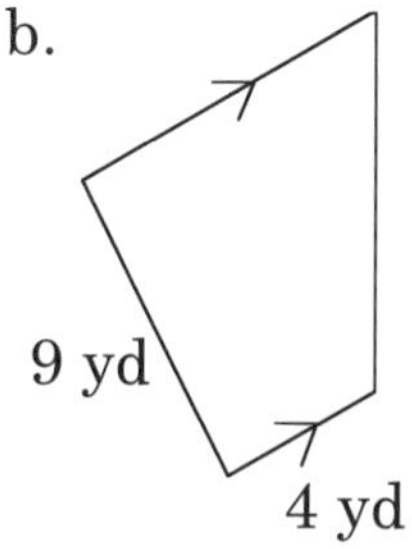

c.
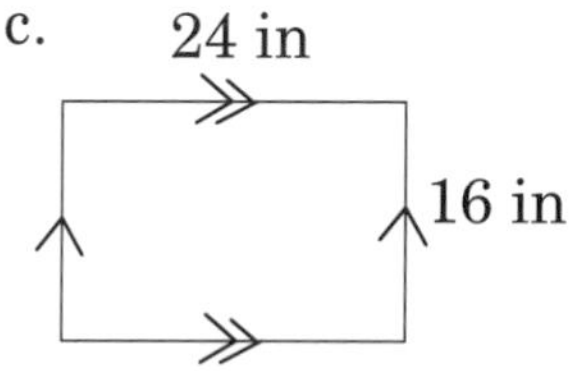

d.
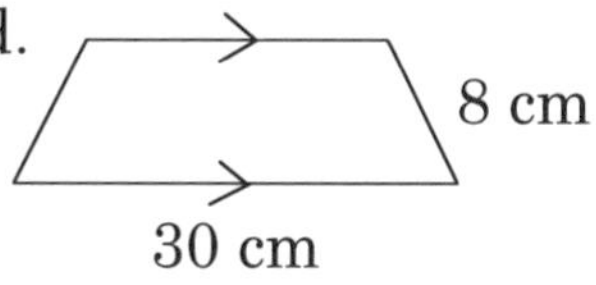

e.
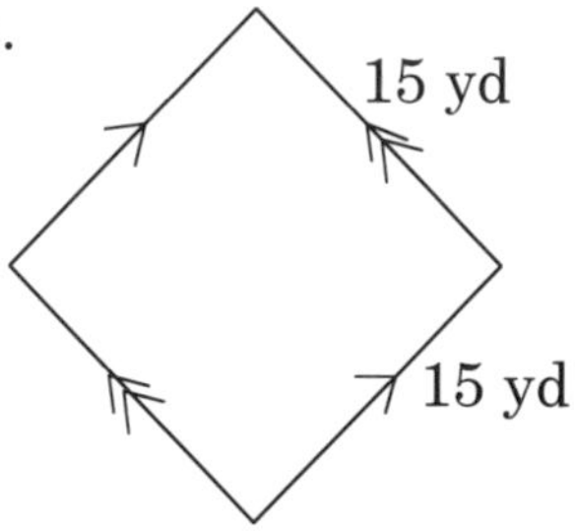

Part 3 Figure out all the missing numbers. Then answer the questions.

This table is supposed to show the number of cars and trucks parked on two lots.

a. Are there more cars or more trucks on lot A?

b. What is the total number of cars on both lots?

c. Which lot has more vehicles in all, lot A or lot B?

d. How many cars are parked on lot B?

e. 285 tells the number of [redacted] on lot [redacted].

	cars	trucks	total for both vehicles
lot A	96	157	[redacted]
lot B	[redacted]	[redacted]	347
total for both lots	158	[redacted]	[redacted]

Part 4

- An angle is formed when two lines **come together.**
- The angle is marked as part of a circle between the lines.

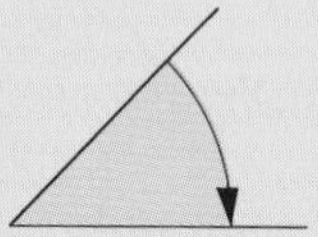

- We can measure angles in units called **degrees.**
  - Here's an angle of 5 degrees:

  - Here's how to write 5 degrees: 5°
- An angle with more than 5 degrees shows a bigger part of the circle. Here's an angle of 45 degrees:

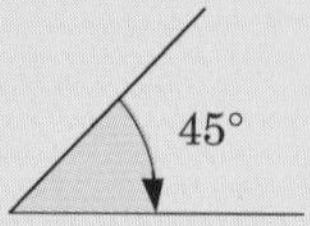

- Each of these angles is 45 degrees:

a. 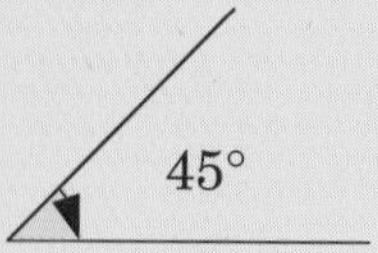

b. 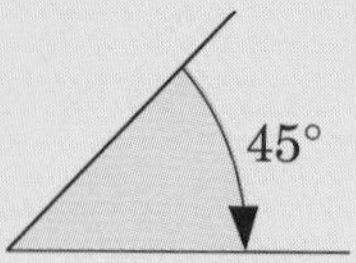

- The same angle can be shown in different positions. All these angles are 45 degrees:

Part 5

Here are important facts about angles:

- The angle for the corner of a room or the corner of a rectangle is 90 degrees.

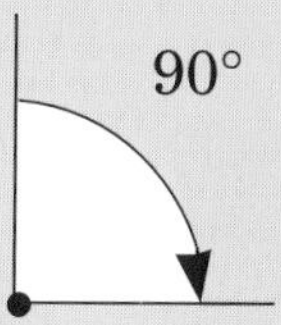

- The angle for a half a circle is 180 degrees. A straight line forms a 180° angle.

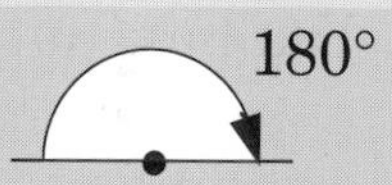

- The angle for a complete circle is 360 degrees.

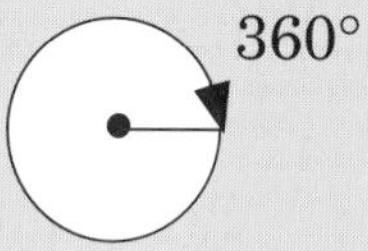

Part 6 **Answer each item.**

a. Write the degrees for the largest angle.

b. Write the letter of each angle that is smaller than angle P.

c. How many degrees are in a whole circle?

d. How many degrees are in half a circle?

e. What's the letter of the smallest angle?

Part 7 **For each item, make a number family and answer the question.**

a. Building C is 56 feet taller than building D. Building D is 194 feet tall. How tall is building C?

b. A truck went 231 miles farther than a van went. If the truck went 721 miles, how far did the van go?

c. Pile A weighed 134 pounds less than pile B. Pile A weighed 198 pounds. How much did pile B weigh?

d. Richard has $95 more than Sarah in his savings account. If Richard has $342, how much does Sarah have?

Part 8

You have learned to write the remainder for division problems. The remainder can also be written as a fraction. When you write the remainder as a fraction, you show the answer as a mixed number. **You change the remainder into a fraction by dividing.**

- The answer to this problem is 4 with a remainder of 6.
- The problem divides by 7. So you can divide the remainder by 7. That's $\frac{6}{7}$.

$$7\overline{)34} = 4\text{ r}6 \qquad 7\overline{)34} = 4\tfrac{6}{7}$$

- The answer to this problem has a remainder of 4.
- The problem divides by 9. So you can divide the remainder by 9. That's $\frac{4}{9}$.

$$9\overline{)31} = 3\text{ r}4 \qquad 9\overline{)31} = 3\tfrac{4}{9}$$

- The answer to this problem has a remainder of 1.
- The problem divides by 5. So you can divide the remainder by 5. That's $\frac{1}{5}$.

$$5\overline{)16} = 3\text{ r}1 \qquad 5\overline{)16} = 3\tfrac{1}{5}$$

Remember, when you have a remainder, you can write it as a fraction. **The number you divide by tells the denominator for the fraction.**

Part 9 Rewrite the complete answer to each problem as a mixed number.

Sample problem $\begin{array}{r} 3 \text{ r4} \\ 7\overline{)25} \end{array}$

a. $\begin{array}{r} 6 \text{ r3} \\ 5\overline{)33} \end{array}$ b. $\begin{array}{r} 2 \text{ r3} \\ 4\overline{)11} \end{array}$ c. $\begin{array}{r} 7 \text{ r1} \\ 3\overline{)22} \end{array}$ d. $\begin{array}{r} 6 \text{ r5} \\ 6\overline{)41} \end{array}$

Independent Work

Part 10 For each problem, write a ratio equation. Then answer the question.

a. There are 12 boxes in every 3 cartons. If there are 96 boxes, how many cartons are there?

b. The ratio of boys to girls in Kelly School is 6 to 5. There are 370 girls in Kelly School. How many boys attend Kelly?

c. In a building, there were 4 windows that were dirty for every 9 windows that were clean. There were 819 clean windows. How many were dirty?

Part 11 For each item, make a number family. Answer the question the problem asks.

a. A lot had 350 cars on it. 49 of those cars were blue. How many were not blue?

b. A kennel had 57 hounds and 85 dogs that were not hounds. How many dogs were in the kennel?

c. 71 of the workers are women. There is a total of 148 workers. How many are men?

Part 12 Copy the table and complete all the rows.

	Fraction equation	Multiplication	Division
a.		$7 \times \blacksquare = 91$	
b.	$\frac{201}{3} = \blacksquare$		
c.			$16\overline{)448}$
d.		$4 \times \blacksquare = 216$	

Part 13 Rewrite each equation so it begins with the letter. Below, write what the letter equals.

a. 15 x 9 = J

b. 124 – 88 = P

c. 472 + 427 = T

Part 14 Work each item. Remember the unit name.

a. Each book weighs 24 ounces. There are 9 books on the shelf. What is the total weight of those books?

b. A pie weighed 36 ounces. The pie was divided into 4 pieces that were the same weight. How much did each piece weigh?

c. A plot had 240 acres. It was divided equally among 12 people. How many acres did each person get?

Part 15 Work each item.

a. $\frac{2}{5} = \frac{■}{100}$

b. $7002 - 1341 = ■$

c. $6 \times 327 = ■$

d. $1 + \frac{18}{5} = ■$

e. $\frac{6}{5}\left(\frac{11}{4}\right) = ■$

f. $\frac{7}{9} = \frac{■}{63}$

g. $■ = \frac{12}{4}$

h. $12 = \frac{■}{2}$

i. $\frac{15}{9} - 1 = ■$

j. $■ + 47 = 47$

k. $1 \times ■ = 91$

l. $72 - ■ = 1$

m. $\frac{4}{8} + \frac{8}{8} = ■$

n. $\frac{2}{8}\left(■\right) = \frac{16}{16}$

o. $\frac{■}{5} = 20$

p. $\frac{20}{4} = ■$

Part 16 Figure out the area of each rectangle. Complete A = ■.

a.

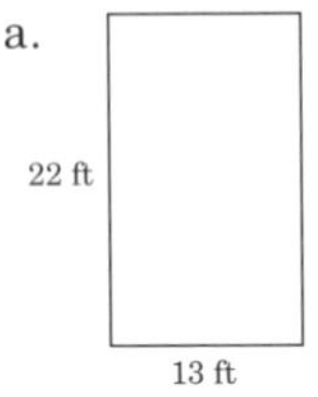

b.

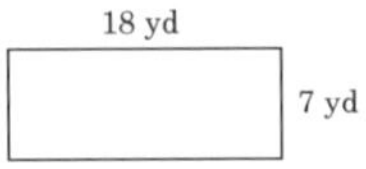

Part J

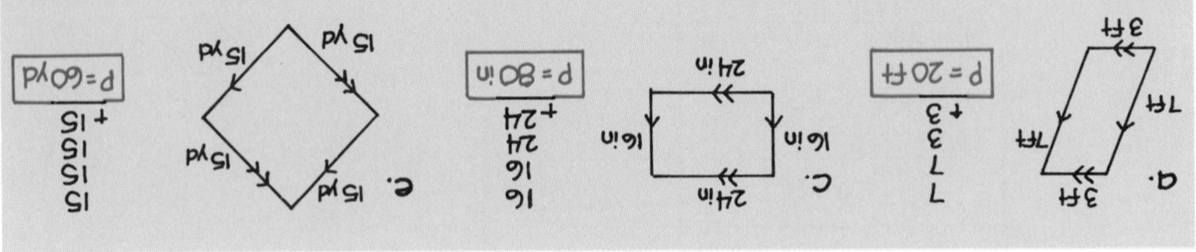

Lesson 24

Part 1 **Copy each table. Use the rules to write the answer for each row.**

a. Multiply each number by 5 and add 4.

3	19
7	■
9	■
5	■

b. Multiply each number by 9 and subtract 2.

6	■
2	■
8	■
1	■
9	■

Part 2 **Rewrite the answer to each problem as a mixed number.**

a. $\begin{array}{r} 9\text{ r1} \\ 3\overline{)28} \end{array}$ b. $\begin{array}{r} 4\text{ r3} \\ 7\overline{)31} \end{array}$ c. $\begin{array}{r} 6\text{ r1} \\ 2\overline{)13} \end{array}$

d. $\begin{array}{r} 6\text{ r3} \\ 4\overline{)27} \end{array}$ e. $\begin{array}{r} 4\text{ r4} \\ 5\overline{)24} \end{array}$ f. $\begin{array}{r} 2\text{ r5} \\ 9\overline{)23} \end{array}$

I guess you could call those mixed numbers, but I don't think they are the type the teacher wants.

Part 3 Make the complete table. Then answer the questions.

This table is supposed to show the number of visitors in two parks during the months of July and August.

Facts

1. 304 people visited Mountain Park in August.
2. 221 people visited Mountain Park in July.
3. The number of people who visited the parks in August was 403.
4. During July and August, a total of 216 people visited Valley Park.

	Valley Park	Mountain Park	total for both parks
July	[redacted]	[redacted]	[redacted]
August	[redacted]	[redacted]	[redacted]
total for both months	[redacted]	[redacted]	[redacted]

Questions

a. Did Mountain Park have more visitors in July or August?

b. 99 tells about the number of visitors in [redacted] during [redacted].

c. What was the total number of visitors for July?

d. What was the total number of visitors for both months?

e. 525 tells about the number of visitors in [redacted] during [redacted].

Part 4 Trace figures that have two pairs of parallel sides. Find the perimeter of those figure.

a.
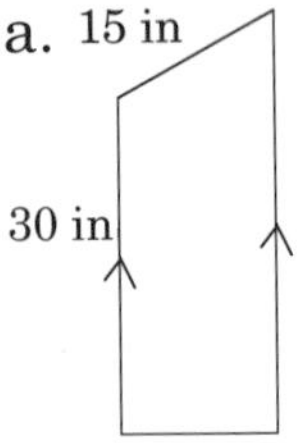

b.
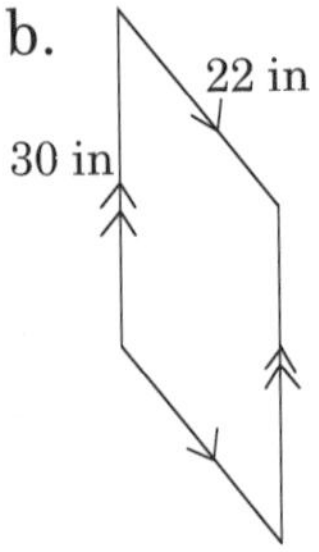

c.
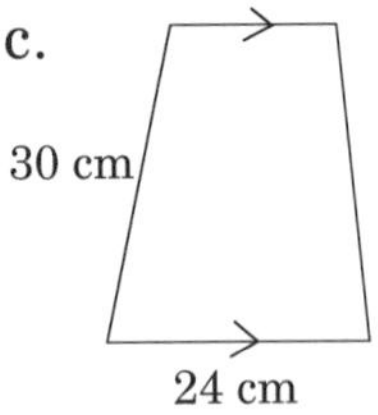

d.
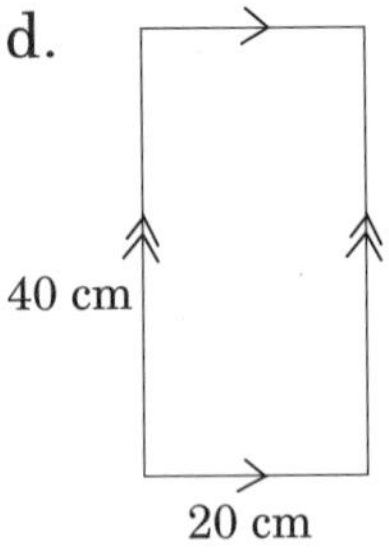

Part 5 **For each item, make a number family and answer the question.**

a. Blaze Lake is 54 miles shorter than Scott Lake. If Scott Lake is 143 miles long, how long is Blaze Lake?

b. There are 15 fewer vacation days in the winter than in the summer. If there are 41 vacation days in the winter, how many vacation days are there in the summer?

c. Sally weighed 32 pounds more in February than she weighed in November. If she weighed 120 pounds in November, what did she weigh in February?

d. Simon has 88 more baseball cards in his collection than Peter. If Simon has 205 cards, how many cards does Peter have?

Part 6 **Answer each question.**

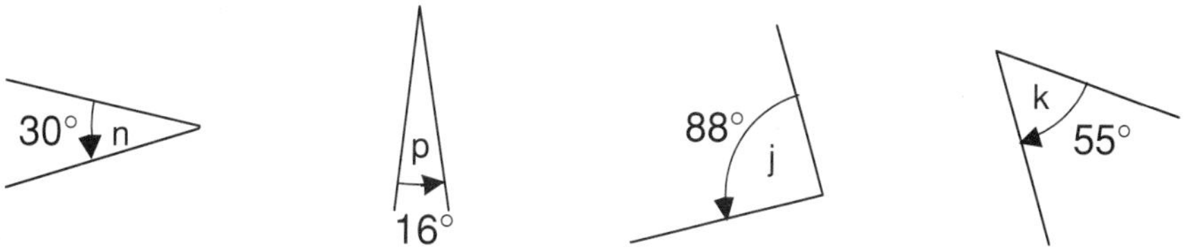

a. Which angle is the largest?
How large is that angle?

b. What is the sum of angles N and P?

c. How many degrees are in the smallest angle?
What's the letter of the smallest angle?

d. How much larger is angle J than angle K?

Independent Work

Part 7 **For each item, write a column problem and the answer with a unit name.**

a. There were 12 eggs in each carton. A store had 60 eggs. How many cartons did the store have?

b. A store has 45 cases of popcorn. Each case weighs 78 pounds. How many pounds of popcorn does the store have?

c. A store divided 98 pies into groups that were the same size. If the store made 7 groups, how many pies were in each group?

Part 8 **Work each item.**

a. [redacted] $= \frac{8}{8}$

b. $\frac{5}{3}$([redacted]) $= \frac{25}{30}$

c. [redacted] $- 10 = 10$

d. $47 +$ [redacted] $= 48$

e. $\frac{\text{[redacted]}}{10} = 5$

f. $\frac{8}{5} = \frac{88}{\text{[redacted]}}$

g. $\frac{12}{7} + 1 =$ [redacted]

h. $307 + 12 + 1001 =$ [redacted]

i. $\frac{3}{7} = \frac{\text{[redacted]}}{42}$

j. $23 \times 49 =$ [redacted]

k. $\frac{4}{7}\left(\frac{9}{9}\right) =$ [redacted]

l. $\frac{36}{4} =$ [redacted]

m. $\frac{28}{10} + 1 =$ [redacted]

n. [redacted] $\times 23 = 0$

o. $\frac{1}{5} - \frac{1}{5} =$ [redacted]

p. $8 = \frac{\text{[redacted]}}{2}$

Part 9 **Write an equation to figure out what to multiply the first fraction by to get the second fraction. If the fractions shown are equivalent, write the simple equation below.**

a. $\frac{4}{9}, \frac{16}{36}$

b. $\frac{12}{5}, \frac{240}{70}$

c. $\frac{8}{3}, \frac{56}{21}$

d. $\frac{9}{7}, \frac{90}{63}$

Part 10 **For each item, write the names and a complete equation.**

a. Part of a road was gravel. Part was paved. There were 3 feet of paved road for every 7 feet of gravel road. If there were 693 feet of paved road, how many feet of the road was gravel?

b. The ratio of ducks to geese in a game preserve is 4 to 3. There are 448 ducks in the preserve. How many geese are there?

c. In a kennel, there were 4 hungry dogs for every 5 dogs that were not hungry. The kennel had 40 dogs that were not hungry. How many dogs were hungry?

Part 11 **Figure the area of each rectangle Complete A = ______.**

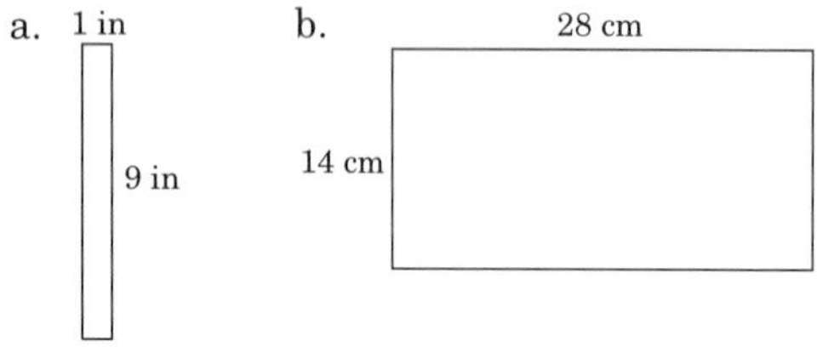

Part 12 **For each item, make a number family. Answer the question the problem asks.**

a. A store had 25 pairs of sneakers. The rest of the shoes in the store were not sneakers. The store had a total of 288 pairs of shoes. How many pairs of shoes were not sneakers?

b. There are 29 puppies and 76 adult dogs in a kennel. How many dogs are in the kennel?

c. A building had 384 lights. 149 of these lights were turned on. How many lights were off?

Part J

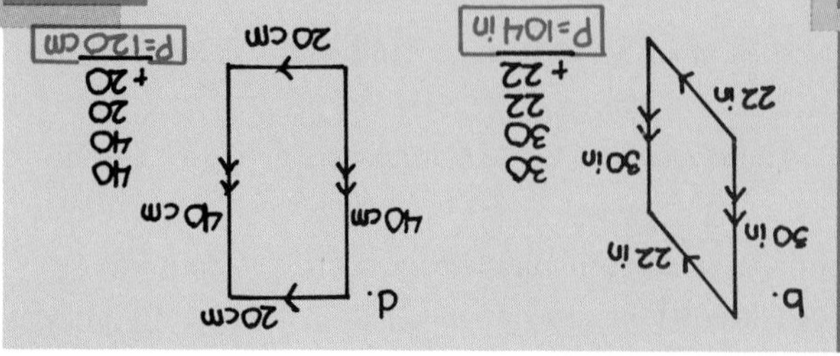

Part K

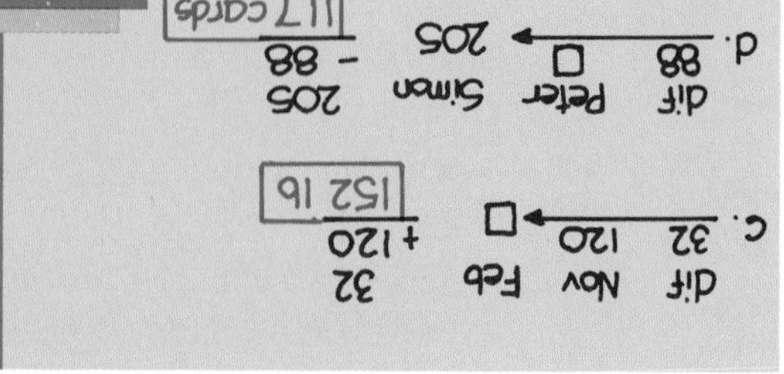

Lesson 25

Part 1 **For each item, make a number family and answer the question.**

a. There are 128 more students in school A than in school B. There are 439 students in school B. How many students are in school A?

b. There were 165 fewer residents in Bloom Town than in Forest Town. There were 829 residents in Bloom Town. How many residents were in Forest Town?

c. The skateboard was 48 pounds lighter than the wagon. The wagon weighed 56 pounds. How much did the skateboard weigh?

d. It is 540 miles farther to Blue River than to Cool Lake. It is 964 miles to Blue River. How far is it to Cool Lake?

Part 2 **Copy and work each problem. Write each answer as a mixed number.**

a. $2\overline{)821}$ b. $4\overline{)823}$ c. $3\overline{)23}$

d. $5\overline{)43}$ e. $6\overline{)121}$ f. $7\overline{)79}$

Part 3 **For each item, make a number family with three letters. Then work each problem.**

a. 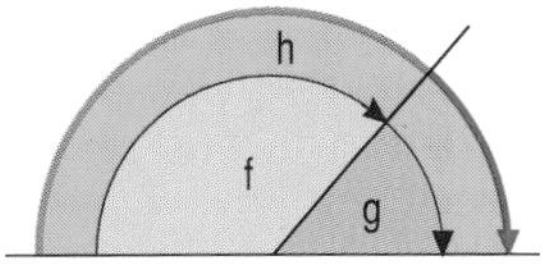

Angle h is $\frac{1}{2}$ a circle. Angle f is 131°. How large is angle g?

b. 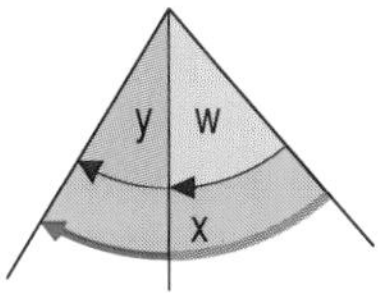

Angle y is 32°. Angle w is 41°. Figure out angle x.

c. 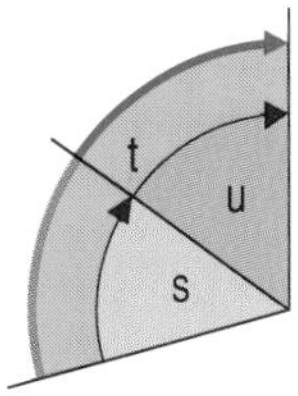

Angle t is 105°. Angle s is 50°. How large is angle u ?

d. 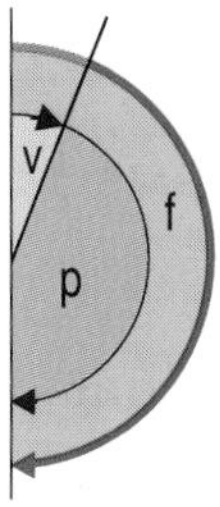

Angle v is 22°. Figure out angle p.

Part 4

Find the perimeter and area of each rectangle.

a.

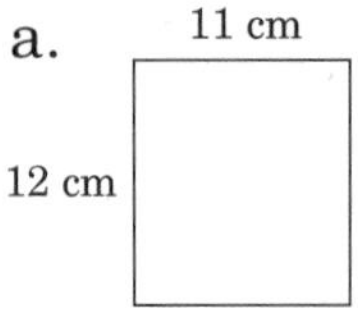

b.

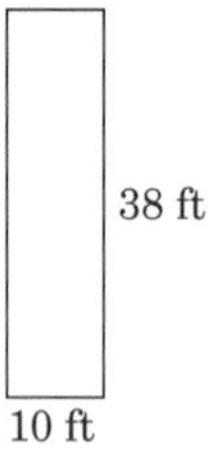

Part 5

Make a table for each item. Follow the rule and write the answers.

a. Multiply each number by 8. Then add 7.

3, 9, 7, 6

b. Multiply each number by 6. Then add 6.

5, 8, 4, 0

Part 6

This table is suppose to show the number of children and adults at two different camps.

Facts

1. There were 17 children at Ho Camp.
2. There were 11 adults at Fun Camp.
3. The total number of people at Fun Camp was 107.
4. The total number of adults for both camps was 226.

	children	adults	people
Ho Camp	[redacted]	[redacted]	[redacted]
Fun Camp	[redacted]	[redacted]	[redacted]
total for both camps	[redacted]	[redacted]	[redacted]

Questions

a. Did Fun Camp have more children or adults?

b. What was the total number of people for both the camps?

c. Which camp had fewer children, Ho Camp or Fun Camp?

d. What was the total number of people in Ho Camp?

e. 11 tells about the number of [redacted] in [redacted].

Independent Work

Part 7

For each item, write the equation showing the fraction and the whole number it equals.

a. The denominator of this fraction is 7 and the fraction equals 6.

b. The denominator of this fraction is 8 and the fraction equals 8.

c. The denominator of this fraction is 1 and the fraction equals 8.

Part 8

For each pair of lines write parallel, not parallel or intersect.

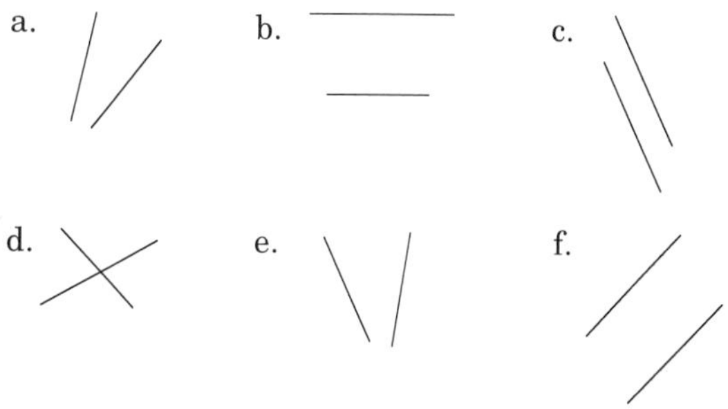

Part 9

Write the answer to each question. Remember the degree symbol in your answer.

a. How many degrees are in a half circle?

b. How many degrees are in a whole circle?

c. How many degrees are in a corner angle?

Part 10

For each item, make a number family. Answer the question the problem asks.

a. There were 89 cars on a long bridge. 34 of the cars were not moving. The rest were moving. How many were moving?

b. 34 of the dogs in a kennel had collars. The rest of the dogs didn't have collars. If 18 dogs did not have collars, how many dogs were in the kennel?

c. A pie was made with filling and crust. The filling weighed 28 ounces. The crust weighed 11 ounces. What was the total weight of the pie?

Part 11

Copy and complete each equation.

a. $\blacksquare = \frac{42}{7}$ b. $12 = \frac{\blacksquare}{3}$

c. $\blacksquare = \frac{12}{3}$ d. $3 = \frac{\blacksquare}{12}$

Part 12

Copy and complete each equation.

a. $\frac{5}{9} = \frac{45}{\blacksquare}$ b. $\frac{3}{2} = \frac{\blacksquare}{60}$

c. $\frac{4}{5} = \frac{\blacksquare}{100}$ d. $\frac{1}{8} = \frac{40}{\blacksquare}$

Part 13

For each problem, write the names and a complete equation. Then box your answer with a unit name.

a. On a farm, the ratio of mice to cats is 5 to 2. If there are 45 mice on the farm, how many cats are there?

b. A machine cuts 3 yards of material every 2 seconds. The machine cuts for 30 seconds. How much material does it cut during this period?

c. A truck drops off 90 pounds of sand every 9 seconds. How many pounds of sand does the truck drop off in 45 seconds?

d. The ratio of tadpoles to frogs in a pond is 9 to 2. If there are 288 tadpoles, how many frogs are in the pond?

Part 14 **Rewrite each equation so it begins with the letter. Below, write what the letter equals.**

a. 87 x 12 = B

b. 418 – 189 = T

c. 11 x 55 = J

d. 820 + 46 + 210 = R

Part J

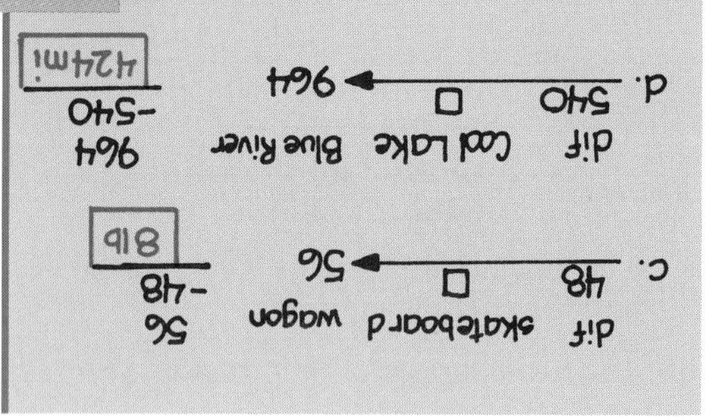

Lesson 26

Part 1

The simplest way to solve some problems is to make a table.

> **There were red cars and black cars on two lots. The lots are Al's lot and Lisa's lot. There were 12 black cars on Lisa's lot. The total number of red cars was 98. Al's lot had 15 red cars. The total number of cars for both lots was 176.**
>
> a. How many black cars were on Al's lot?
>
> b. What's the total number of black cars for both lots?
>
> c. Did Lisa's lot have more red cars or black cars?

- The first names in the problem are **column** headings.
- There were red cars and black cars on two lots.
- The column headings are **red cars, black cars** and **total.**

	red cars	black cars	total

- The next names in the problem are **row** headings.
- The lots are Al's lot and Lisa's lot.
- The row headings are **Al's lot, Lisa's lot** and **total.**

	red cars	black cars	total
Al's lot			
Lisa's lot			
total			

- Remember, the **first** names in the problem are the **column** headings. The **next** names are the **row** headings.
- Now you can read the rest of the problem, fill in the missing numbers and answer the questions.

Part 2 **Make a table for the problem. Answer the questions.**

Game wardens put trout and bass in two different lakes. The lakes were Blue Lake and Ross Lake. The wardens put 125 trout in Blue Lake. They put 74 bass in Blue Lake and 128 bass in Ross Lake. They put a total of 231 fish in Ross Lake.

a. How many trout were put in Ross Lake?

b. How many total fish were put in Blue Lake?

c. What's the total number of bass for both lakes?

Part 3 **Write each fraction as a division problem and work it. Write the answer as a mixed number.**

a. $\frac{18}{7}$ b. $\frac{25}{3}$ c. $\frac{44}{6}$ d. $\frac{37}{4}$ e. $\frac{23}{7}$

Part 4 **For each item, make a number family with three letters and two numbers. Then figure out the angle.**

a.

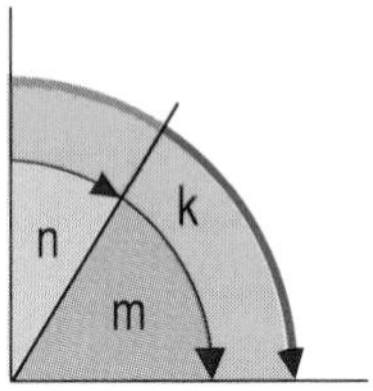

Angle m is 58°. Angle k is 90°. Figure out angle n.

b.

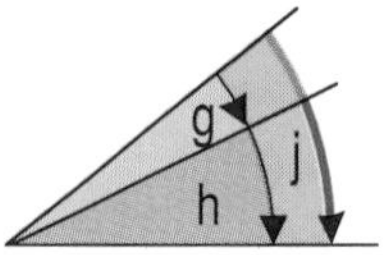

Angle h is 25°. Angle j is 38°. Figure out angle g.

c.

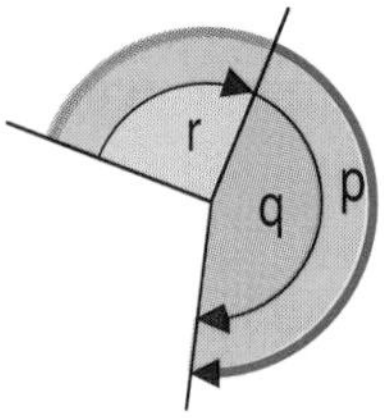

Angle p is 254°. Angle r is 90°. Figure out angle q.

d.

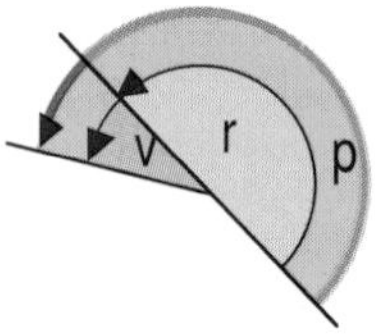

Angle v is 33°. Angle r is $\frac{1}{2}$ a circle. Figure out angle p.

Part 5 **Find the perimeter and area of each rectangle.**

a.

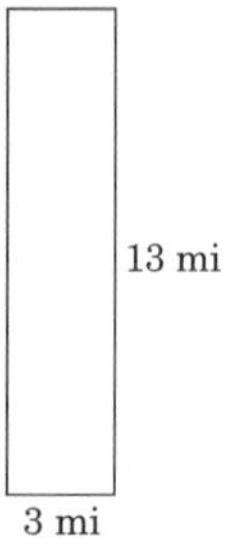

b.

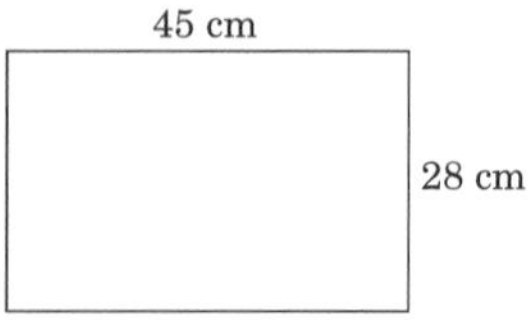

c.

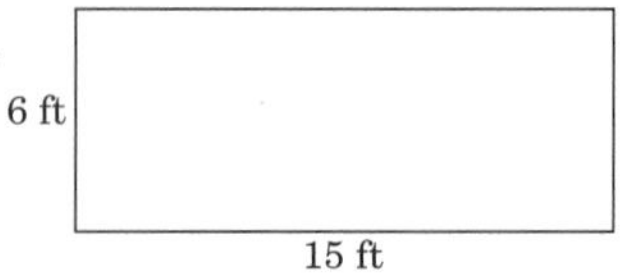

Part 6 **Work each item.**

a. 145 more students attend the middle school than the high school. There are 540 students at the middle school. How many students attend the high school?

b. There are 155 fewer cherry trees in the orchard than apple trees. There are 49 cherry trees. How many apple trees are there?

c. Phyllis is 34 years younger than her father. If her father is 50 years old, how old is Phyllis?

Part 7 **Make a table for each item. Follow the rule and write the answers.**

a. Multiply each number by 4. Then subtract the product from 37.
6, 5, 3, 2

b. Multiply each number by 7. Subtract the product from 47.
6, 4, 5, 2

Independent Work

Part 8 **Write the letter of each pair of lines that are not parallel. Write intersect or do not intersect for each of those pairs.**

Part 9 **Work each item.**

a. $1 - \frac{13}{14} =$

b. $\frac{6}{1} \times \frac{5}{3} =$

c. $\frac{4}{5} + \frac{8}{5} =$

d. $\frac{16}{3} \times \frac{1}{1} =$

Part 10 **For each problem, write the names and a complete equation.**

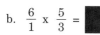

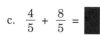

a. There are 12 cans in every 2 packages. How many packages are needed for 288 cans?

b. The ratio of buds to flowers in a rose garden is 7 to 3. If there are 77 buds, how many flowers are there?

c. A large vehicle needs 3 gallons of gas to go 8 miles. How many gallons would the vehicle use in 560 miles?

d. At a picnic, there were 9 flies for every 2 people. There were 136 people at the picnic. How many flies were there?

Part 11 For each item, make a number family. Answer the question the problem asks.

a. 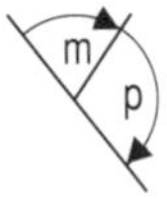

Angle p is 117°. How many degrees is angle m?

b. 102 of the vehicles on a lot were cars. The rest were vans. If there was a total of 167 vehicles on the lot, how many were vans?

c. In a laundry, 197 articles of clothing were wet. The rest were dry. If there were 642 articles of clothing in the laundry, how many were dry?

Part 12 Some of the angles are more than 180°. Some are less than 180°. Write the letter of each angle that is more than 180°.

a. 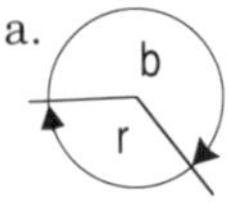

b. 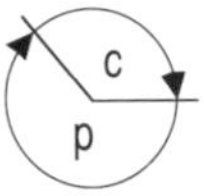

c. 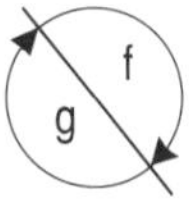

d. 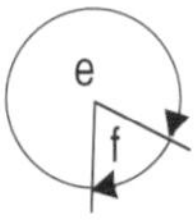

e. 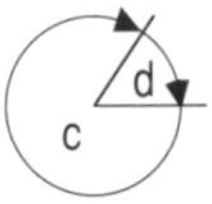

Part 13 Copy and complete each equation.

a. $\blacksquare = \frac{27}{3}$ b. $\frac{\blacksquare}{8} = 3$ c. $6 = \frac{\blacksquare}{5}$ d. $130 = \frac{\blacksquare}{5}$ e. $\blacksquare = \frac{264}{8}$

Part 14 Find the perimeter of each figure.

a. 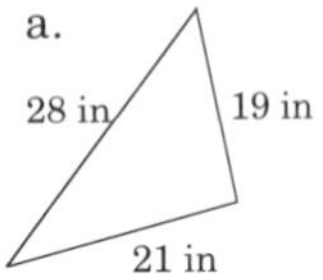

b. 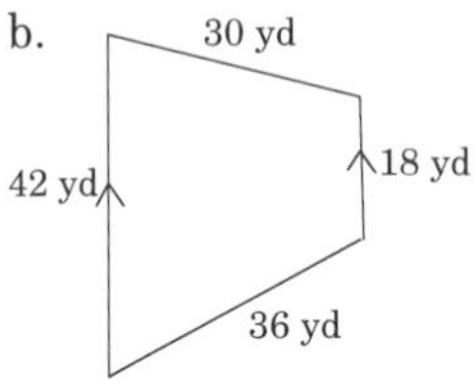

c. 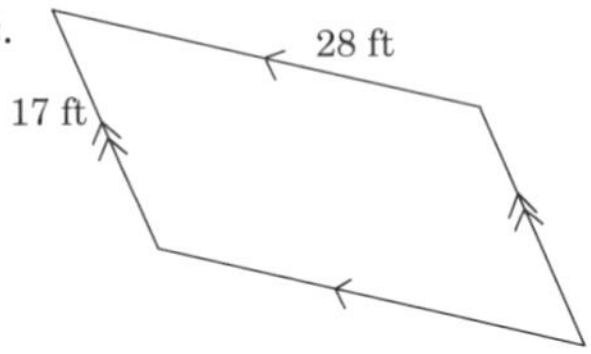

Part 15 Rewrite each equation so it begins with the letter. Below, write what the letter equals.

a. 26 x 11 = P b. 960 – 488 = J c. 560 + 210 + 27 = T d. 22 x 20 = K

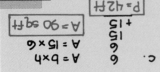

Part J

a. 13
13
3
+ 3
P = 32 mi

A = b × h
A = 3 × 13
A = 39 sq mi

b. 28
28
45
+ 45
P = 146 cm

A = b × h
A = 45 × 28
A = 1260 sq cm

c. 6
6
15
+ 15
P = 42 ft

A = b × h
A = 15 × 6
A = 90 sq ft

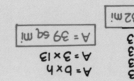

Lesson 27

Part 1

- If two lines form a 90° angle, the lines are **perpendicular.** Perpendicular edges are found in lots of things that people build.
- The walls of a room are perpendicular to the ceiling.

- The floor of a room is perpendicular to the walls.

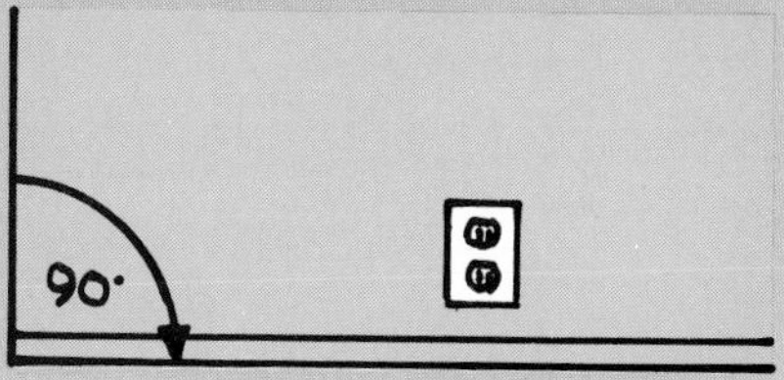

- The top and bottom of your paper are perpendicular to the sides.

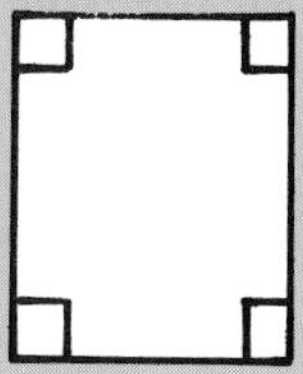

- Perpendicular lines are marked with a special angle marker. The marker is shaped like a square corner to show that the angle is 90°.

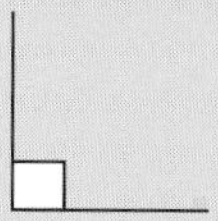

Part 2

Trace the figure. Figure out all the angles that are marked with letters.

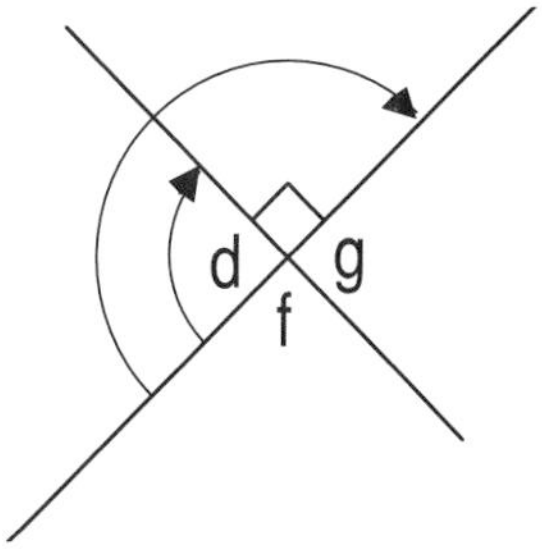

Part 3

Write each fraction as a division problem and work it. Write the answer as a mixed number or a whole number.

a. $\frac{31}{3}$ b. $\frac{127}{4}$ c. $\frac{95}{5}$ d. $\frac{163}{6}$ e. $\frac{462}{7}$

Part 4

- A line that intersects parallel lines creates the same pair of angles at each parallel line.
- Here's a pair of parallel lines:

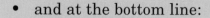

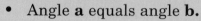

- Here's a line intersecting both parallel lines:

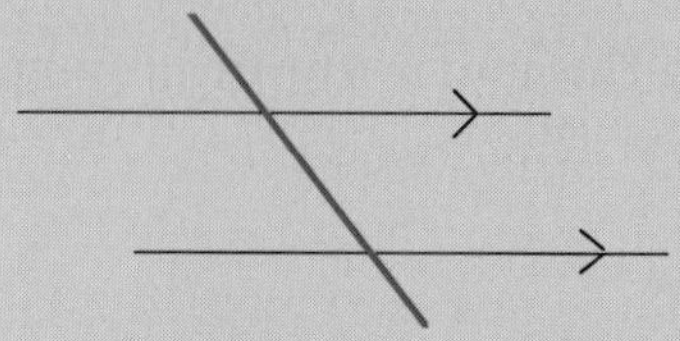

- The intersecting line creates the same angle at the top line:
- and at the bottom line:

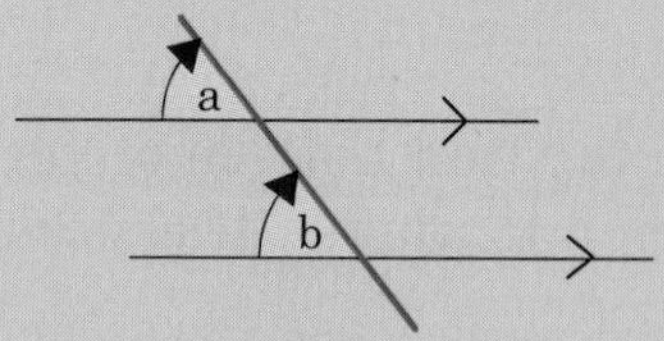

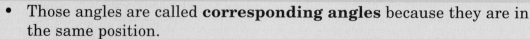

- Angle **a** equals angle **b.**
- Those angles are called **corresponding angles** because they are in the same position.
- Here's angle **m** at the top line and corresponding angle **n** at the bottom line:
- They're equal to each other.

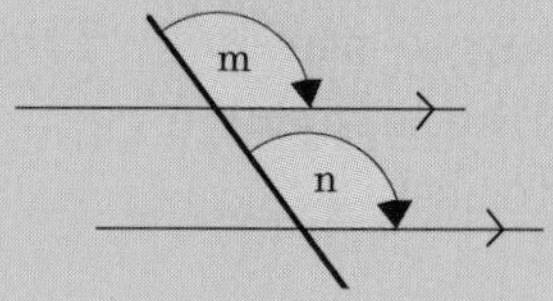

- Here's another pair of parallel lines and an intersecting line:

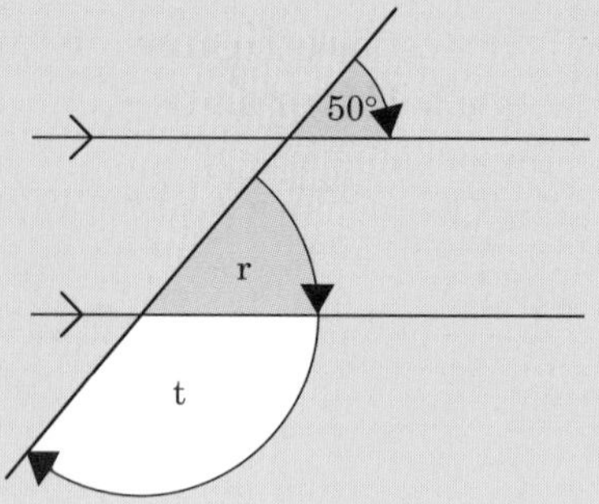

- The angle at the top parallel line is 50°. So the corresponding angle at the bottom line is 50°. That's angle **r.**
- If you know that angle **r** is 50°, you can figure out angle **t.** Angles **r** and **t** make half a circle. So angle **t** is: 180° – 50°. That's 130°.

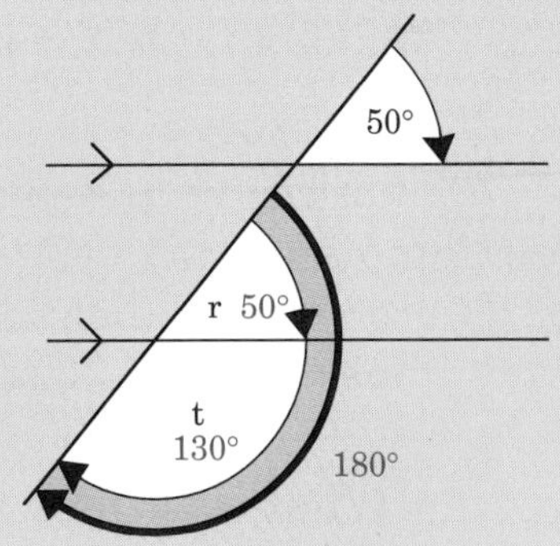

Part 5 **Figure out all the angles marked with a letter.**

a.

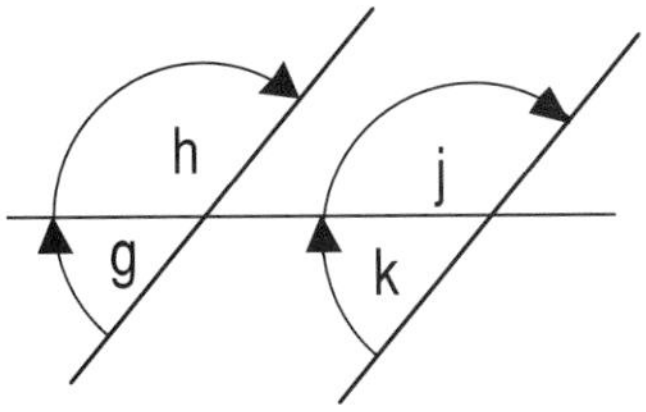

Two lines are parallel.
Angle g = 50°.

b.

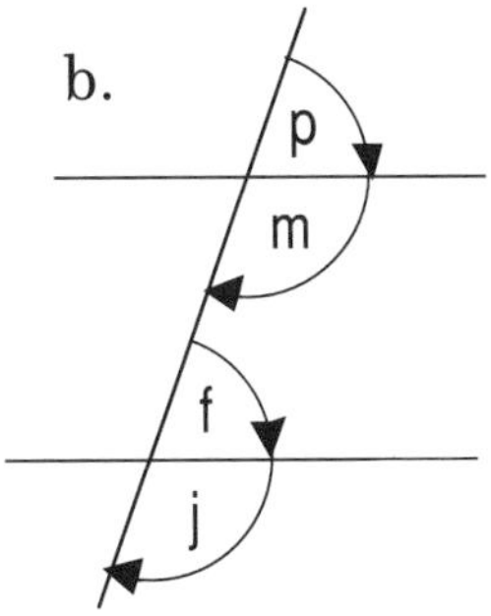

Two lines are parallel.
Angle f = 70°.

Part 6 **Work each item.**

a. The ratio of red flowers to yellow flowers is 2 to 7. If there are 10 red flowers, how many yellow flowers are there?

b. There are 8 bees for every 3 flowers. If there are 99 flowers, how many bees are there?

c. Tom eats 3 ounces of food for every 2 ounces his little sister eats. Tom eats a total of 69 ounces of food. How much does his sister eat?

d. The ratio of sand to gravel was 5 pounds to 6 pounds. If the mixture had 240 pounds of gravel, how much sand was in the mixture?

Part 7 Make a table for the problem. Answer the questions.

There were both boys and girls at two different camps. The camps were Rainier Camp and Maxwell Camp. There were 517 boys at Maxwell Camp. There were 98 girls at Maxwell Camp. The total number of boys for both camps was 642. The total number of children for both camps was 1213.

a. What was the total number of children at Maxwell Camp?

b. Were there more boys or girls at Rainier Camp?

c. Which of the two camps had fewer girls?

d. 125 tells about the number of [redacted] at [redacted].

Part 8 Make a table for each item. Follow the rule and write the answers.

a. Add 14 to each number. Then divide the sum by 3.
10, 4, 1, 7

b. Multiply each number by 32. Then subtract the product from 255.
7, 2, 5

Independent Work

Part 9 For each problem, write the names and a complete equation.

a. The ratio of men to women at a dance is 5 to 6. If there are 120 women, how many men are there?

b. It takes 6 workers to carry 2 beams. How many beams could 54 workers carry?

Part 10 For each item, write a column problem and the answer with a unit name.

a. 900 pounds of sand is divided into 30 piles that are the same weight. How many pounds are in each pile?

b. Each carton holds 6 eggs. If there are 900 eggs, how many cartons are needed?

c. A machine caps 36 bottles each minute. How many bottles does the machine cap in 18 minutes?

d. 256 biscuits are placed in packages that each hold 8 biscuits. How many packages are needed?

Part 11 **Figure out the perimeter and area for each rectangle.**

a.

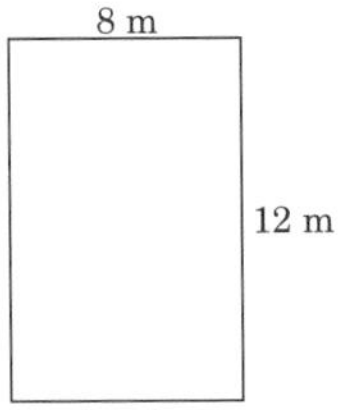

b.

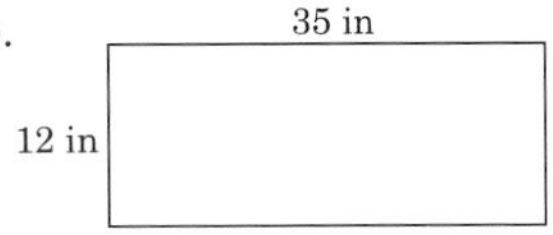

Part 12 **Work each item.**

a. $\frac{14}{3} - \frac{14}{3} = \blacksquare$

b. $1 + \frac{8}{2} = \blacksquare$

c. $\frac{4}{9}(\blacksquare) = \frac{20}{36}$

d. $\frac{6}{3} \times \frac{8}{3} = \blacksquare$

e. $\frac{5}{4} - 1 = \blacksquare$

f. $\frac{\blacksquare}{8} = 8$

g. $\frac{11}{3} - \frac{8}{3} = \blacksquare$

h. $\frac{14}{2} = \blacksquare$

i. $1 + \frac{5}{5} = \blacksquare$

Part 13 **Work each item.**

a. In the morning, Tina ran 874 meters. In the afternoon, she ran 937 meters farther than she did in the morning. How far did Tina run in the afternoon?

b. Carl had 112 more stamps than Frieda had. Carl had 327 stamps. How many stamps did Frieda have?

c. In a lake, there were 38 fewer motorboats than sailboats. There were 38 motorboats in the lake. How many sailboats were there?

Part 14 **Write a multiplication equation for each item. If the fractions shown are equivalent, write the simple equation below.**

a. $\frac{1}{8}, \frac{8}{48}$ b. $\frac{3}{5}, \frac{27}{40}$ c. $\frac{7}{2}, \frac{21}{6}$

Part 15 **Write the equation to show the fraction each mixed number equals.**

a. $5\frac{9}{15}$ b. $2\frac{11}{12}$

c. $206\frac{1}{2}$ d. $11\frac{2}{3}$

Part 16 **For each item, make a number family and answer the question.**

a.

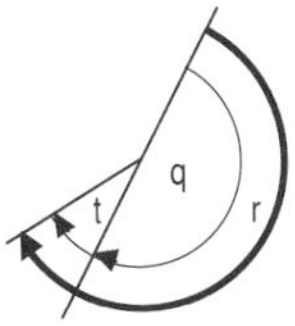

Angle t is 32°. How many degrees is angle r?

b.

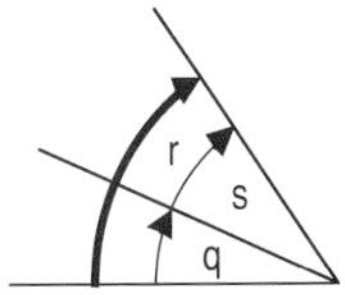

Angle s is 32°. Angle r is 56°. How many degrees is angle q?

c.

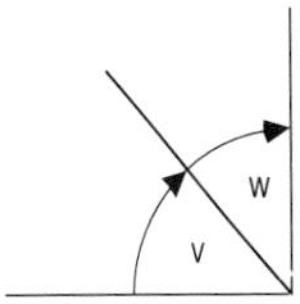

Angle v is 49°. How many degrees is angle w?

Lesson 28

Part 1 **For each pair of lines, write parallel, not parallel or intersect.**

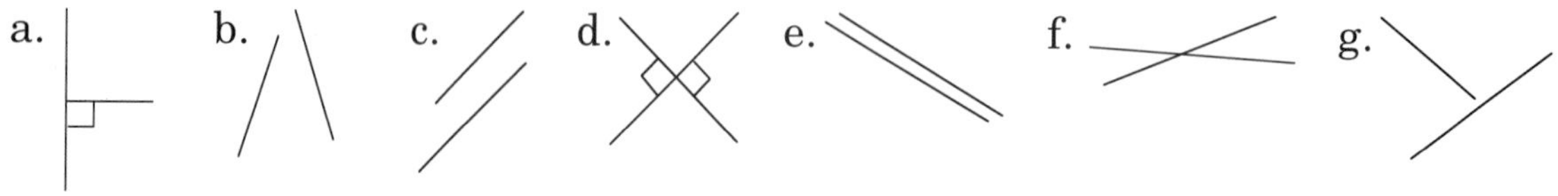

Part 2 **Make a table for the problem. Answer the questions.**

In 1980 and 1981, babies were born in Queen's Hospital and Marist Hospital. In 1980, there were 173 babies born in Queen's Hospital. In 1981, the number of babies born in both hospitals was 575. The number of babies born in Marist Hospital during both years was 463. In 1980, the total number of babies for both hospitals was 237.

a. How many babies were born in Queen's Hospital in 1981?

b. In 1981, were more babies born in Queen's Hospital or Marist Hospital?

c. In which year were fewer babies born in Marist Hospital?

d. 349 tells about the number of babies born in [redacted] during [redacted].

Part 3 **Figure out all the angles marked with a letter.**

a.

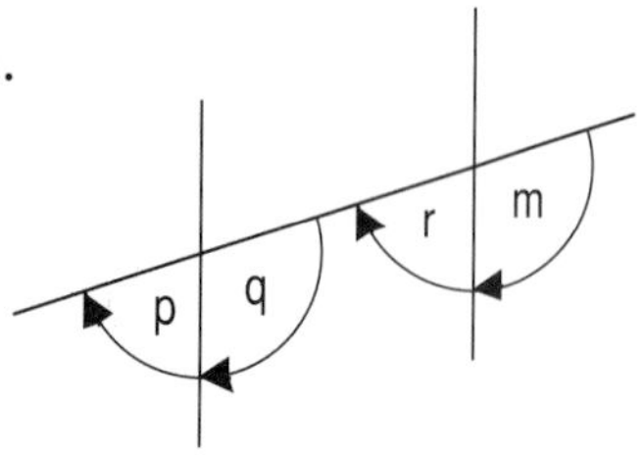

Two lines are parallel.
Angle p is 72°.

b.

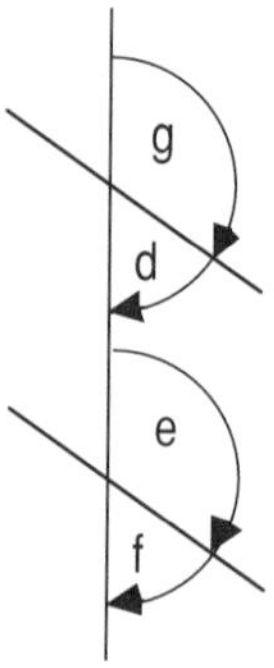

Two lines are parallel.
Angle g is 125°.

Part 4

- You've worked with figures that have pairs of parallel sides. These figures are called **parallelograms.**
- A **rectangle** is a special kind of parallelogram.

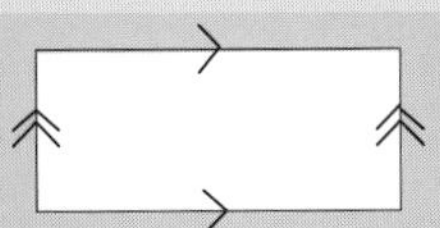

- A **square** is a parallelogram.

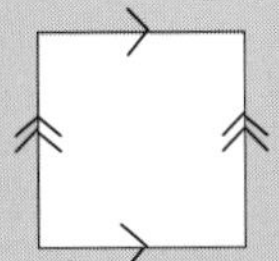

- These figures are also parallelograms.

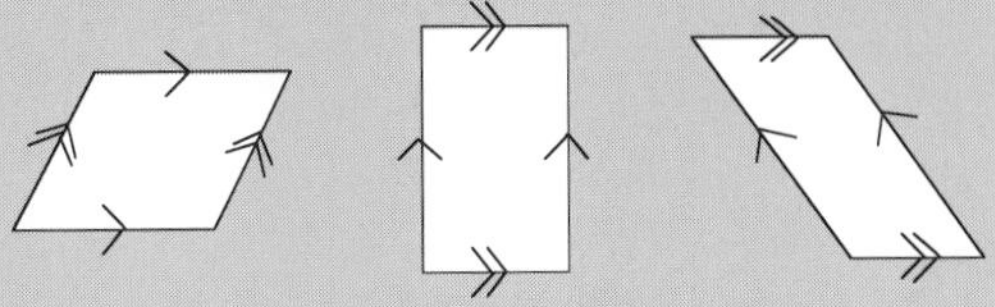

- You can use **the equation for the area of a rectangle** to find the area of any parallelogram.

Area = base x height

- For parallelograms that are not rectangles, the **height** of the figure is **not** the length of a side.
- The left side does not show the height. The **height** is a dotted line that is perpendicular to the base of the figure.

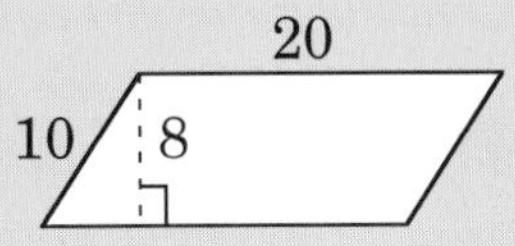

- To find the area for the parallelogram, you work the problem 20 x 8.
- You can make a rectangle that has the same area as the parallelogram.

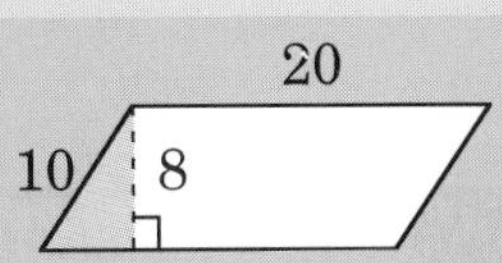

- You do that by moving the shaded triangle that is on the left side to the right side.

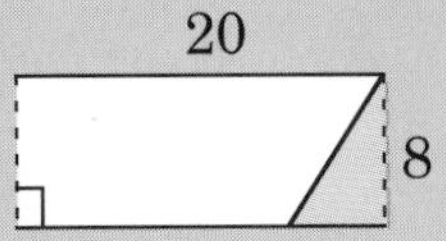

- The parallelogram and the rectangle have the same area.
- Remember, the equation for the area of the parallelogram is **Area = base x height,** but the height may not be the length of a side.

Part 5 Find the area of each figure.

a.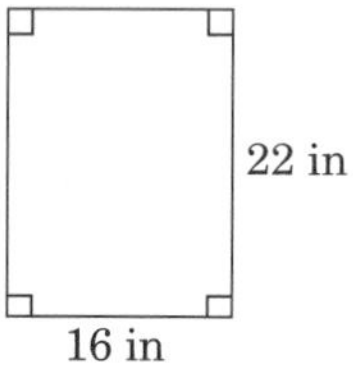
22 in
16 in

b.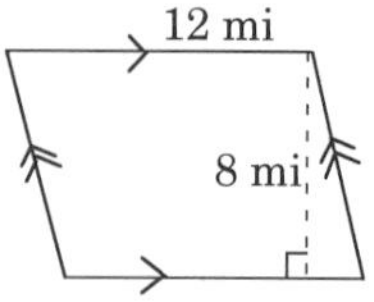
12 mi
8 mi

c.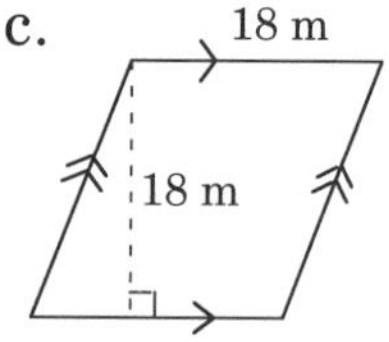
18 m
18 m

d.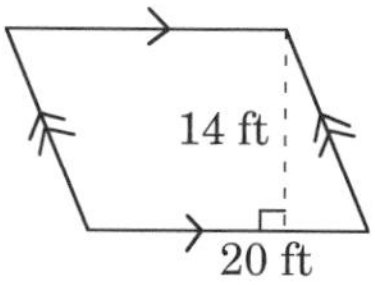
14 ft
20 ft

Part 6 Copy and complete the table.

x 8	
2	■
■	■
total 9	■

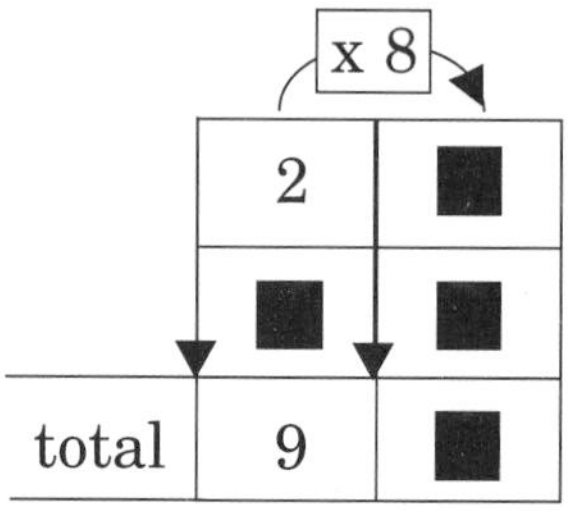

Part 7 Write each fraction as a division problem and work it. Write the answer as a mixed number.

a. $\dfrac{38}{5}$

b. $\dfrac{146}{7}$

c. $\dfrac{27}{2}$

d. $\dfrac{352}{9}$

Independent Work

Part 8 Make a table for each item.

a. Multiply each number by 9. Then add 4.
   2, 8, 5, 9

b. Multiply each number by 16. Then subtract the product from 139.
   3, 8

c. Divide each number by 5. Then add 21.
   5, 50, 0, 20

Part 9 Write a multiplication equation for each item. If the fractions shown are equivalent, write the simple equation below.

a. $\dfrac{8}{9}$, $\dfrac{72}{72}$

b. $\dfrac{4}{7}$, $\dfrac{16}{28}$

c. $\dfrac{5}{3}$, $\dfrac{40}{24}$

Part 10 For each item, write the names and a complete equation. Then answer the question.

a. A truck unloads 36 pounds of gravel every 3 seconds. How many seconds will it take the truck to unload 1800 pounds of gravel?

b. 3 boxes hold 7 pounds of crackers. How many boxes are needed for 567 pounds of crackers?

c. There were 7 eggs for every 5 chickens. There were 50 chickens. How many eggs were there?

Part 11

For each item, make a number family. Answer the question the problem asks.

a. The horse weighed 138 pounds less than the bull. The bull weighed 1678 pounds. How much did the horse weigh?

b. The Cap building is 188 feet tall. The Hop Building is 147 feet shorter than the Cap Building. How tall is the Hop Building?

c. The train traveled 502 miles farther than the bus traveled. The bus traveled 609 miles. How far did the train travel?

d. A turtle is 21 years older than the tree. The turtle is 134 years old. How old is the tree?

Part 12

For each item, write the fraction and the whole number it equals.

a. $■ = \frac{35}{7}$

b. $■ = \frac{200}{4}$

c. $8 = \frac{■}{2}$

d. The fraction equals 3 wholes. The denominator is 10.

e. The fraction equals 80. The denominator is 1.

f. $\frac{■}{20} = 4$

g. $\frac{■}{5} = 20$

Part 13

Copy and complete each table.

a.

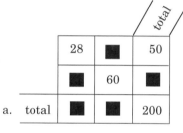

			total
	28	■	50
	■	60	■
total	■	■	200

b.

			total
	■	135	■
	■	■	508
total	■	604	1020

Part 14

Rewrite each equation so it begins with the letter. Below, write what the letter equals.

a. 40,706 + 29,689 = B

b. 3229 – 568 = R

c. 412 x 36 = K

d. $\frac{1256}{8} = T$

Part 15

Copy the table and complete all the rows.

Multiplication	Fraction equation	Division
	$\frac{1020}{30} = ■$	$\overline{)\quad}$
	$\frac{364}{13} = ■$	
4 (■) = 280		

Lesson 29

Part 1 **Make a table for the problem. Answer the questions.**

Hard wheat and soft wheat were grown on different farms–Brown Farm and Roger Farm. Brown Farm produced 680 tons of soft wheat. Roger Farm produced 295 tons of hard wheat. The total amount of wheat produced on Roger Farm was 450 tons. The total amount of wheat produced on both farms was 1265 tons.

a. What is the total amount of hard wheat grown on both farms?

b. Was more hard wheat grown on Brown Farm or Roger Farm?

c. Was there less hard wheat or soft wheat on Roger Farm?

Part 2 **Write each fraction as a division problem and work it. Write the answer as a mixed number.**

a. $\frac{191}{4}$ b. $\frac{35}{8}$ c. $\frac{141}{5}$ d. $\frac{26}{3}$

Part 3 **Find the area of each figure.**

a.

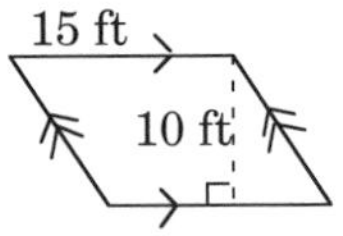

b.

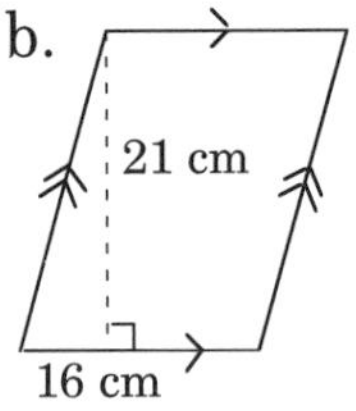

c.

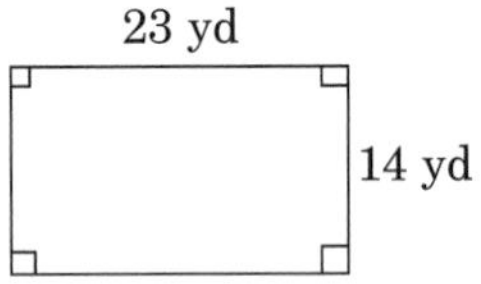

Part 4 **Copy and complete each table.**

x 9

	4	■
	■	■
a. total	7	■

	7	56
	9	■
b. total	16	■

	3	■
	10	40
c. total	13	■

Part 5 Figure out all the angles marked with a letter.

a.

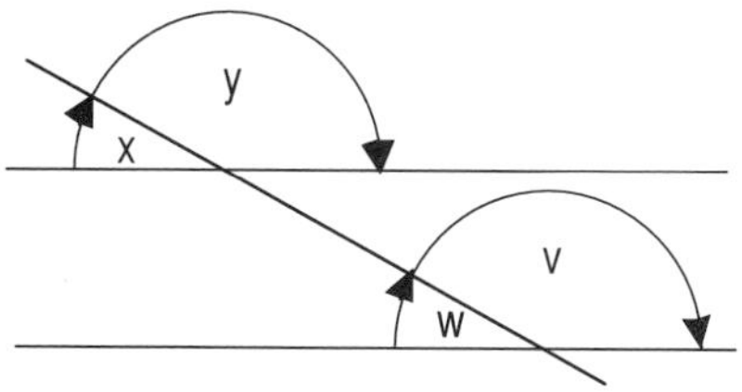

Two lines are parallel.
Angle w = 28°.

b.

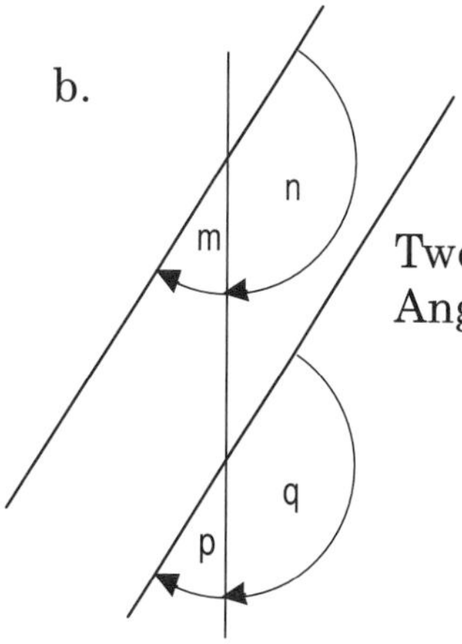

Two lines are parallel.
Angle n = 147°.

Part 6 Make a table for each item. Follow the rule and write the answers.

a. Multiply each number by 0. Then add 4.
2, 8, 5, 9

b. Add 4 to each number. Then multiply the sum by 5.
3, 1, 8

c. Multiply each number by 20. Then subtract the product from 100.
5, 2, 3

Independent Work

Part 7 For each item, write parallel, not parallel or intersect. If intersecting lines are perpendicular, also write perpendicular.

a. b. c. d. e. f. g.

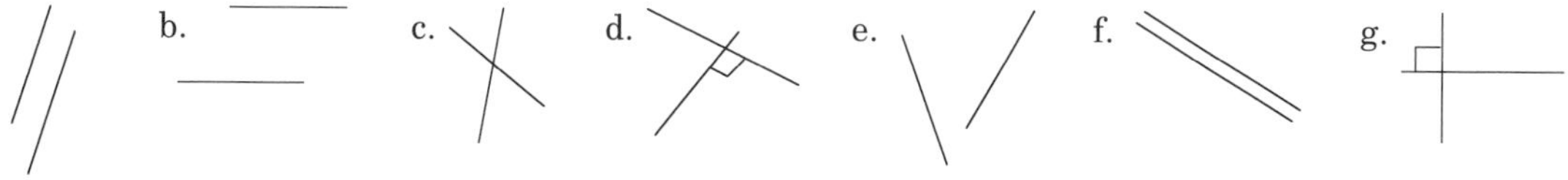

Part 8 For each item, write the fraction and the whole number it equals.

a. $■ = \frac{96}{12}$ b. $4 = \frac{■}{1}$ c. $2 = \frac{■}{300}$ d. $80 = \frac{■}{3}$ e. $■ = \frac{208}{4}$

Part 9 **For each item, make a number family. Answer the question each problem asks.**

a. The dog was 18 months younger than the cat. If the cat was 53 months old, how old was the dog?

b. A trailer weighed 4880 pounds. The trailer weighed 2260 pounds less than a car. How much did the car weigh?

c. In a garden, there were 71 more tomatoes than potatoes. There were 468 potatoes. How many tomatoes were there?

d. Jane had read 19 fewer books than her brother, Greg. Greg had read 46 books. How many had Jane read?

Part 10 **For each item, make a number family and answer the question.**

a.

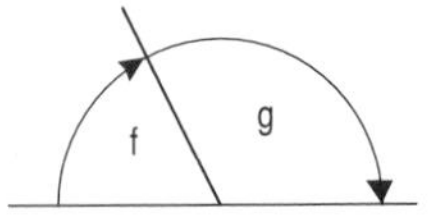

Angle g is 117°.
How many degrees is angle f?

b.

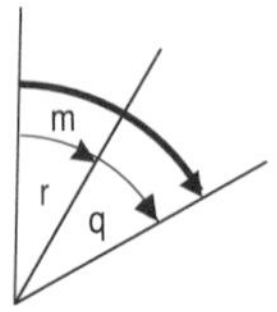

Angle q is 31°. Angle m is 60°.
How many degrees is angle r?

Part 11 **Rewrite each equation so it begins with the letter. Below, write what the letter equals.**

a. $876 - 300 = K$ b. $48 \times 13 = J$ c. $\frac{56}{56} = B$ d. $118 \times 20 = T$

Part 12 **For each problem, write the names and a complete equation. Then answer the question.**

a. There were 4 white marbles for every 9 yellow marbles. There were 80 white marbles. How many yellow marbles were there?

b. A fast-food place sold 2 hamburgers for $3. The store sold a total of 824 burgers. How many dollars did the store receive for these sales?

c. The ratio of water to sand in a mixture is 2 to 5. If there are 200 gallons of water in the mixture, how many gallons of sand are in the mixture?

d. 6 machines cut 8 patterns each minute. How many machines would be needed to cut 48 patterns each minute?

e. The ratio of buds to flowers on a tree was 8 to 7. There were 64 buds on the tree. How many flowers were on the tree?

Part 13 **Find the perimeter of each figure.**

a.

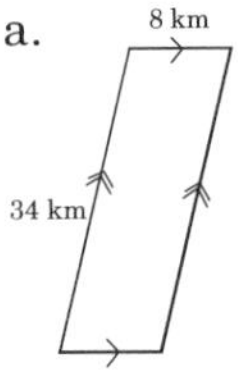

b.

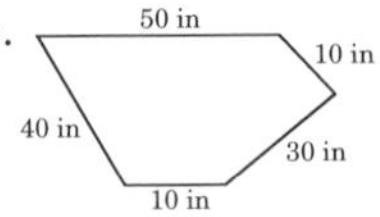

c.

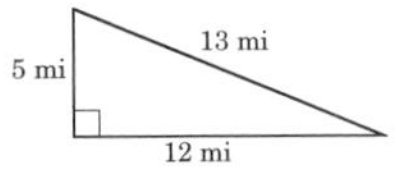

Part 14 **Work each equation.**

a. $\frac{28}{3} - \frac{24}{3} =$ [redacted]

b. $1 - \frac{7}{8} =$ [redacted]

c. $\frac{26}{2} \times \frac{3}{1} =$ [redacted]

d. $\frac{4}{5} \times \frac{1}{2} =$ [redacted]

e. $\frac{3}{8} + \frac{21}{8} =$ [redacted]

f. $\frac{14}{3} - 1 =$ [redacted]

Lesson 30

Part 1 **Copy and work rows a through d.**

	Division	Fraction equation
Sample	$5\overline{)23}$ → $4\frac{3}{5}$	$\frac{23}{5} = 4\frac{3}{5}$
a.	$9\overline{)39}$	
b.	$8\overline{)71}$	
c.	$3\overline{)25}$	
d.	$4\overline{)14}$	

Part 2 **Find the area for each figure.**

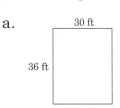

a.

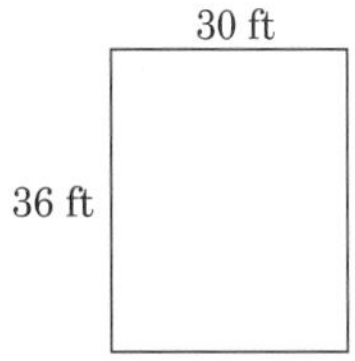

b.

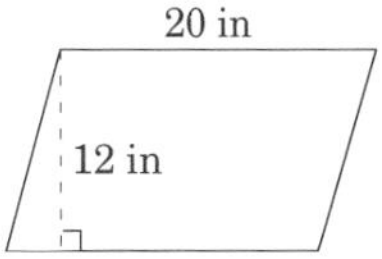

c.

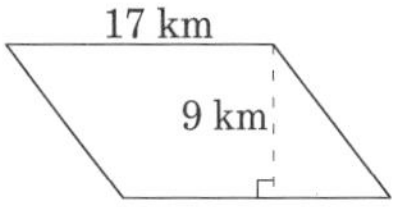

Test 3

Part 1 **For each item, write an equation with the names. Figure out the missing number and answer the question.**

a. In a garden there were 6 worms in every 5 pounds of dirt. A person dug up 84 worms. How many pounds of dirt did the person dig up?

b. The ratio of owls to sparrows in a field is 2 to 11. There are 6 owls in the field. How many sparrows are there?

c. A tractor moved 3 yards of dirt every 7 seconds. The tractor worked for 126 seconds. How many yards of dirt did the tractor move?

Part 2 **Answer each question.**

The table shows the number of oak trees and pine trees in two parks.

a. How many pines are in Hillside Park?

b. How many trees of both types are in Terrace Park?

c. What's the total number of oaks for both the parks?

d. 55 tells about the number of [redacted] trees in [redacted].

e. Which is more, the total number of pines or the total number of oaks?

	oaks	pines	total for both types
Terrace Park	88	146	234
Hillside Park	51	55	106
total for both parks	139	201	340

Part 3 **Work each item.**

a. How many degrees are in half a circle?

b. How many degrees are in a circle?

c. How many degrees are in a corner of a rectangle?

d. When two lines are perpendicular, they form an angle that is [redacted] degrees.

e. Draw a pair of perpendicular lines. Then show the symbol for an angle formed by those lines.

Part 4 **Work each item.**

a. Write the number of degrees in angle k.

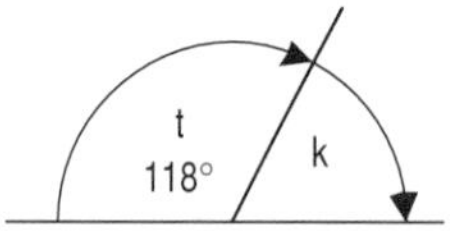

b. Write the number of degrees in angle p.

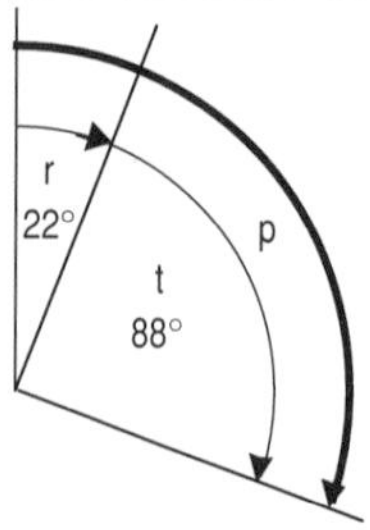

Part 5 **Make a table. Follow the rule and write the answers.**

Multiply by 5. Subtract the product from 79.

3, 8, 2, 0

Part 6

Make a number family for each problem. Answer the questions.

a. The bear cub weighed 340 pounds less than the mother bear. The bear cub weighed 95 pounds. How much did the mother bear weigh?

b. The church is 46 feet taller than the tower. The church is 197 feet tall. How tall is the tower?

c. Ginger collected 194 stamps. Donna collected 306 stamps. How many more stamps did Donna collect than Ginger collected?

Part 7

For each item, write parallel if the lines are parallel. Write intersect if the lines intersect. If lines are not parallel and do not intersect, don't write anything.

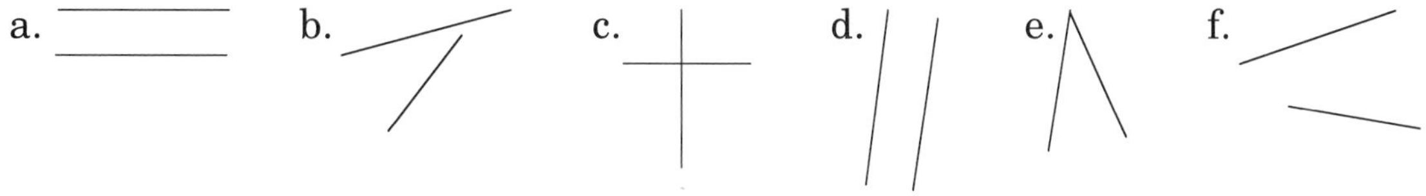

Part 8

Figure the perimeter and the area for each rectangle.

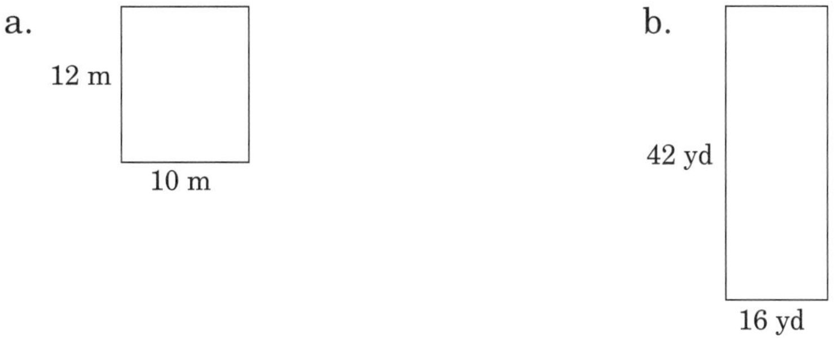

Part J

a. A = b × h
A = 30 × 36
A = 1080 sq ft

b. A = b × h
A = 20 × 12
A = 240 sq in

c. A = b × h
A = 17 × 9
A = 153 sq km

Lesson 31

Part 1

Copy each problem. Multiply and subtract to figure out the remainder. Then write the remainder as a fraction.

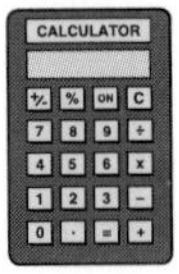

a. $\begin{array}{r} 6 \\ 47\overline{)289} \end{array}$ b. $\begin{array}{r} 5 \\ 36\overline{)205} \end{array}$ c. $\begin{array}{r} 9 \\ 23\overline{)227} \end{array}$ d. $\begin{array}{r} 8 \\ 16\overline{)139} \end{array}$

Part 2

Copy and complete each table.

a.

	4	28
	5	■
total	■	■

b.

	3	■
	1	■
total	■	44

Part 3

Read each number as hundreds.

a. 4500 b. 3600 c. 1100 d. 1800 e. 9200

Part 4

Trace each figure and write the number of degrees for each angle.

a. Angle h = 50°.

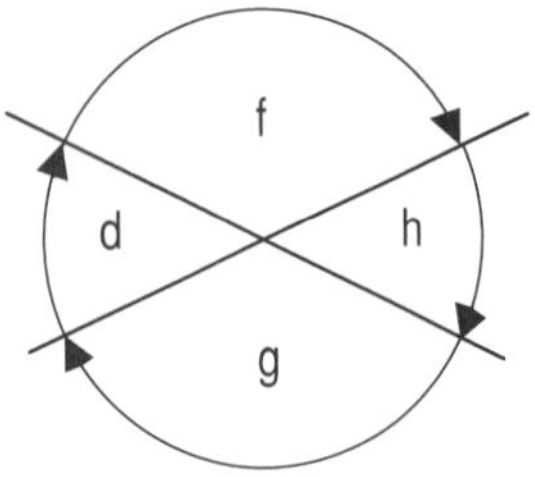

b. Angle m = 115°.

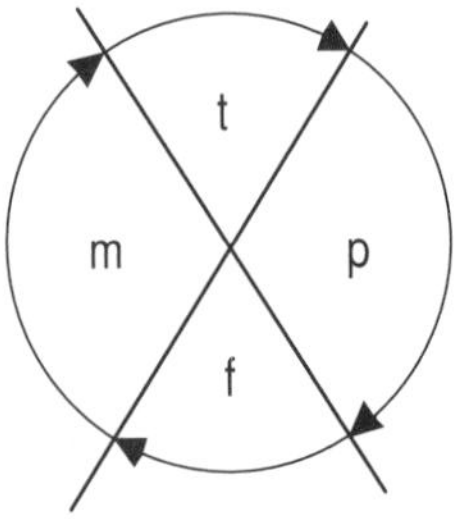

Part 5 **Make a table for the problem. Answer the questions.**

Workers picked apples in September and in October. The apples were Granny Smith apples and Red Delicious apples. Workers picked a total of 1170 bushels of Granny Smith apples. In September, workers picked 276 bushels of Granny Smith apples. In September, workers picked 191 bushels of Red Delicious apples. The total number of apples picked during both months was 3356.

a. How many bushels of Granny Smith apples were picked in October?

b. What was the total number of bushels picked in October?

c. How many bushels of Red Delicious apples were picked in all?

Part 6

- Any fraction that is more than 1 can be written as a whole number or a mixed number.
- You just read the fraction as a division problem, write it as a division problem, and figure out the answer as a mixed number or a whole number.

- Here's: $\frac{14}{9}$
- The fraction is more than 1, so we can write it as a mixed number or a whole number.
- Here it is as a division problem: $9\overline{)14}$
- The answer is $1\frac{5}{9}$. $9\overline{)14}$ with quotient $1\frac{5}{9}$
- So: $\frac{14}{9} = 1\frac{5}{9}$

- Here's: $\frac{36}{9}$
- The fraction is more than 1, so we can write it as a mixed number or a whole number.
- Here it is as a division problem: $9\overline{)36}$
- The answer is 4. $9\overline{)36}$ with quotient 4
- So: $\frac{36}{9} = 4$

Part 7 Copy each fraction that is more than 1. Then write it as a whole number or a mixed number.

a. $\frac{17}{8}$ b. $\frac{13}{15}$ c. $\frac{1}{9}$ d. $\frac{29}{26}$ e. $\frac{63}{7}$ f. $\frac{12}{20}$ g. $\frac{67}{9}$

Independent Work

Part 8 Figure out the area of each parallelogram.

a.
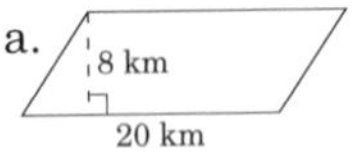

b.
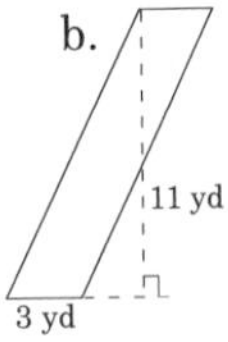

c.
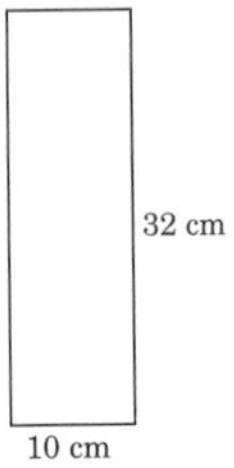

Part 9 Figure out the perimeter of each parallelogram.

a.
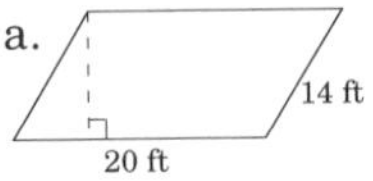

b.
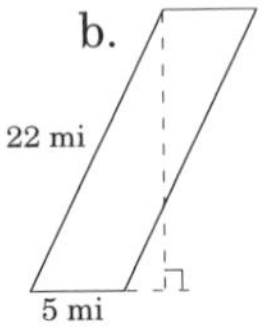

c.
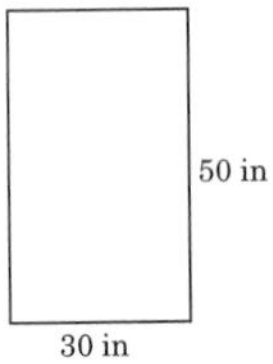

Part 10 For each item, make a number family. Answer the question the problem asks.

a. The lamb weighed 56 pounds less than her mother. The lamb weighed 25 pounds. How much did her mother weigh?

b. The large tent costs $71 more than the small tent. The large tent costs $178. How much does the small tent cost?

c. 114 of the dogs were at Al's Kennel. There were 16 fewer dogs at Dog's World than at Al's Kennel. How many dogs were at Dog's World?

Part 11 Make a table for each item.

a. Add 23 to each number. Divide the sum by 5.
2, 22, 7

b. Multiply each number by 0. Then add 8 to the product.
1, 14, 0, 6

c. Multiply each number by 3. Then add 50.
7, 4, 20, 5

Part 12 For each problem, write the names and a complete equation. Then answer the question.

a. A company charges $8 to clean 9 square yards of carpet. What area of carpet would the company clean for $144?

b. The ratio of cows to horses is 5 to 2. There are 100 horses on the ranch. How many cows are on the ranch?

c. On a lake, there were 4 sailboats for every 7 canoes. If there were 108 sailboats, how many canoes were there?

Part 13 Copy and complete each equation. Below, write a simple equation for the equivalent fractions.

a. $\frac{2}{3}\left(\frac{8}{8}\right) = ■$

b. $\frac{7}{4} = \frac{56}{■}$

c. $\frac{2}{5} = \frac{■}{30}$

d. $\frac{12}{7}\left(\frac{5}{5}\right) = ■$

e. $\frac{1}{7}\left(\frac{30}{30}\right) = ■$

Part 14 Trace the figure. Write the number of degrees for each angle that is marked.

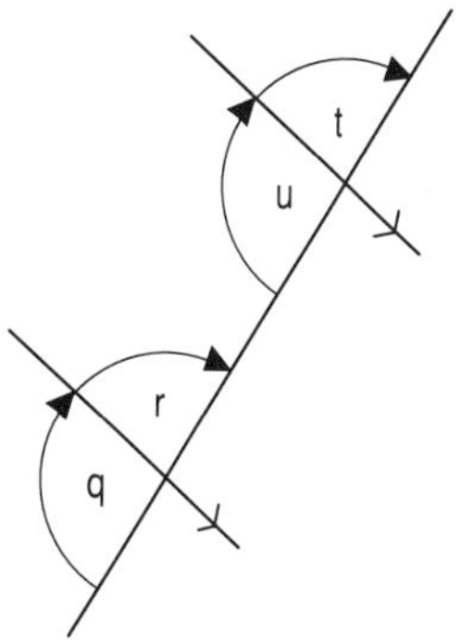

Angle t is 80°.

Part 15 Copy the table and complete all the rows.

	Multiplication	Division	Fraction equation
a.	27 (■) = 108	$\overline{)}$	
b.		■ over 2 $\overline{)560}$	
c.		$\overline{)}$	$\frac{279}{9} = ■$
d.	7 (■) = 154	$\overline{)}$	

Part J

c. 23|227
−207
20
$9\frac{20}{23}$

d. 16|139
−128
11
$8\frac{11}{16}$

Lesson 32

Part 1

Copy each problem. Multiply and subtract to figure out the remainder. Then write the remainder as a fraction.

a. $63\overline{)337}$ with 5 written above

b. $91\overline{)809}$ with 8 written above

c. $49\overline{)358}$ with 7 written above

d. $17\overline{)160}$ with 9 written above

Part 2

Find the area and perimeter of each parallelogram.

a.

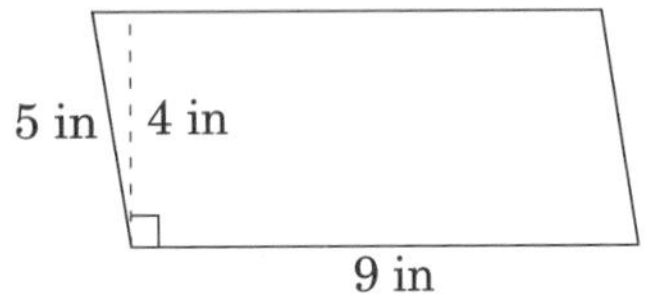

b.

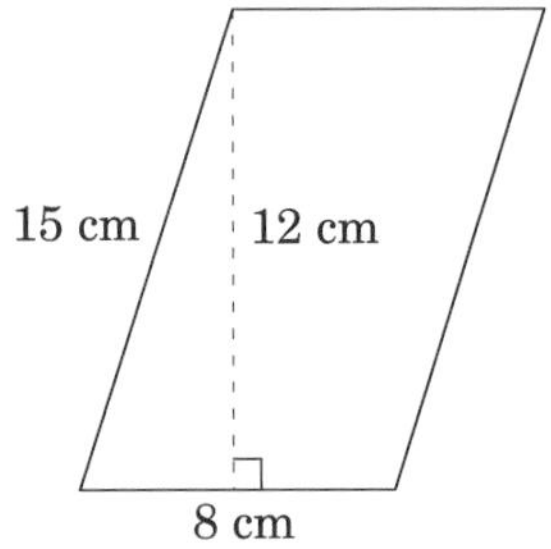

Part 3

- You've worked word problems that compare two things and **tell** the difference number.
- Some problems **ask** about the difference. Questions that ask about the difference name two things and ask which is more or less.
- Here are some questions that ask about the difference:

  How much taller is Jane than Dan?

  How much less money does the car cost than the truck costs?

  What's the difference in price between the basket and the board?

- If a problem **asks** about the difference, make a number family with the name **difference.**

- Here's a problem:

  A TV costs $295. A radio costs $48. What is the difference in the price of these two items?

- The question asks about the difference. So you start with a number family that compares.

  dif
  ————→

- You can figure out the name for the big number by comparing the cost of the TV and the radio. The TV costs more. It's the big number.

- Here's the number family:

  dif **radio** **TV**
  48 ————→ **295**

- The difference number is a small number. To find the difference number, **you always subtract.** The difference is $247. That's how much more the TV costs than the radio costs.

$$\begin{array}{r} \$\,295 \\ -\quad 48 \\ \hline \boxed{\$\,247} \end{array}$$

Part 4 **For each item, make a number family and answer the question. Remember the unit name.**

a. Jenny is 58 inches tall. Tony is 39 inches tall. What is the difference in height?

b. The cable car went 15 miles. The bus went 60 miles. What is the difference in distance?

c. A vat holds 194 gallons. A tank holds 256 gallons. How much more does the tank hold than the vat holds?

Part 5 **Copy and complete each table.**

a.

perch	2	■
bass	■	■
fish	5	40

b.

red	5	■
not red	■	21
cars	8	■

Part 6 Read each number as hundreds.

a. 1300 b. 1324 c. 5606 d. 1128 e. 7210

Part 7 Trace each figure. Then write the number of degrees for each angle.

a. Angle y = 60°.

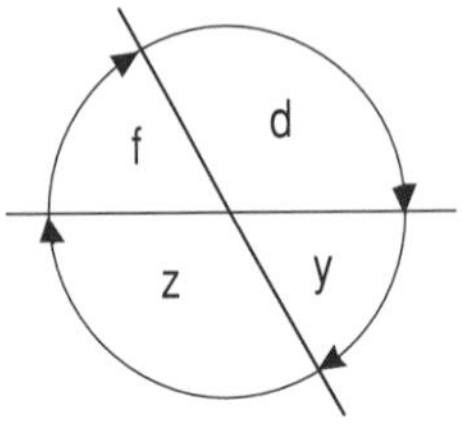

b. Angle t = 135°.

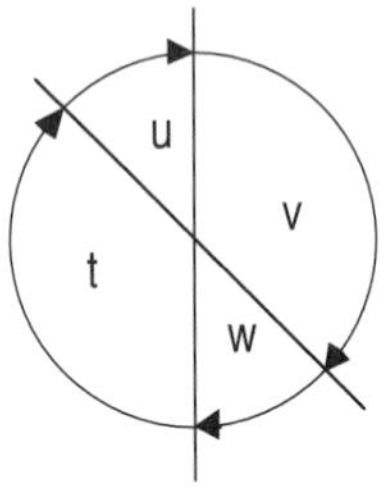

Part 8 Copy each fraction that is more than 1. Then write it as a whole number or a mixed number.

a. $\frac{5}{31}$ b. $\frac{77}{6}$ c. $\frac{17}{20}$ d. $\frac{580}{5}$ e. $\frac{11}{15}$ f. $\frac{18}{17}$ g. $\frac{47}{3}$

Independent Work

Part 9 Here are the two circles you worked with in part 7. There's a rule about angles that are on opposite sides of a circle. Write the rule about opposite angles.

a.

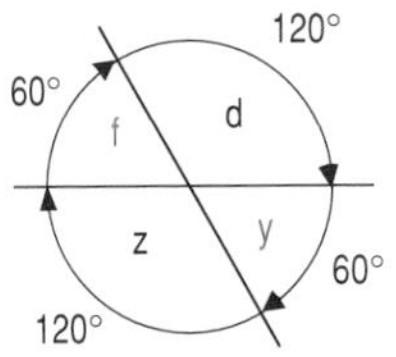

b.

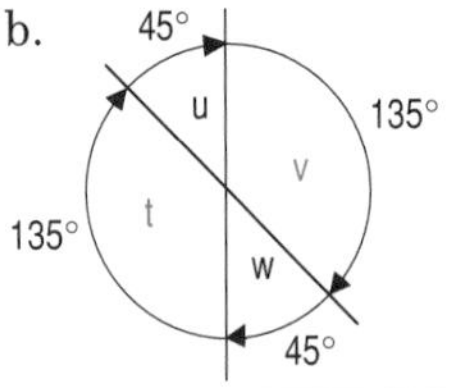

Part 10 Make a table for the problem. Answer the questions.

Two schools served potatoes and other vegetables. The schools were Jackson School and Fisher School. Fisher School served 86 pounds of potatoes and 113 pounds of other vegetables. A total of 160 pounds of potatoes were served at the two schools. A total of 511 pounds of vegetables were served at the two schools.

a. What was the weight of potatoes served at Jackson School?

b. Were more potatoes or other vegetables served at Fisher School?

c. 238 tells about the pounds of [redacted] served at [redacted].

Part 11 **Work each item.**

a. $\frac{3}{4}$ ([redacted]) = $\frac{63}{88}$

b. $\frac{4}{3} \times \frac{6}{3} =$ [redacted]

c. $\frac{3}{7} - \frac{3}{7} =$ [redacted]

d. $8 \times$ [redacted] $= 112$

e. $\frac{8}{5} + \frac{6}{5} - \frac{2}{5} =$ [redacted]

f. $\frac{5}{3} = \frac{\text{[redacted]}}{15}$

g. $\frac{16}{9} - \frac{0}{9} =$ [redacted]

h. $47 -$ [redacted] $= 0$

i. [redacted] $\times 15 = 15$

j. $\frac{1}{5} \times \frac{3}{5} =$ [redacted]

k. $0 \times 264 =$ [redacted]

l. $1 - \frac{11}{15} =$ [redacted]

m. $\frac{7}{3} + 1 =$ [redacted]

Part 12 **Make a table for each item.**

a. Multiply each number by 10. Then subtract the product from 199.
10, 19, 0, 1

b. Subtract 8 from each number. Then multiply the difference by 5.
8, 15, 12, 20

Part 13 **Write the answer for each item.**

a. How many degrees are in a corner of a rectangle?

b. Write the symbol for a 90° angle.

c. How many degrees are in a circle?

d. How many degrees are in half a circle?

Part 14 **Complete each equation. Below, write a simple equation for the equivalent fractions.**

a. $\frac{3}{5}\left(\frac{9}{9}\right) =$ [redacted]

b. $\frac{2}{7} = \frac{\text{[redacted]}}{28}$

c. $\frac{12}{11} = \frac{60}{\text{[redacted]}}$

d. $\frac{3}{8}\left(\frac{60}{60}\right) =$ [redacted]

Part 15 **For each problem, write the names and a complete equation. Then answer the question.**

a. In the ant colony, there were 7 workers for every 9 eggs. The colony had 1260 eggs. How many worker ants were there?

b. A mower could cut 6 square feet of grass in 11 seconds. How many seconds would it take the mower to cut 288 square feet?

c. The ratio of hawks to owls in a forest was 3 to 2. There were 126 owls. How many hawks were in the forest?

d. The ratio of girls to boys in Riverside School is 5 to 6. There are 252 boys in the school. How many girls are there?

Part 16 **Trace the figure. Write the number of degrees for each angle.**

Angle m = 71°.

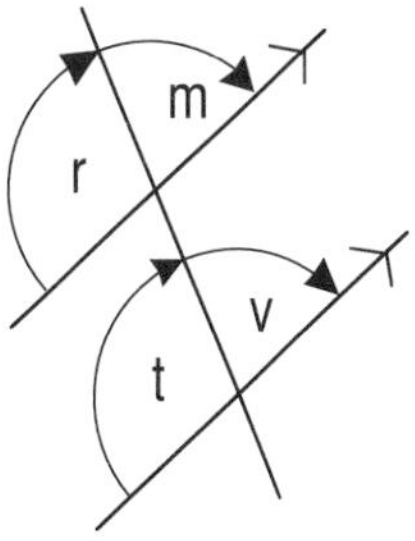

Lesson 33

Part 1

Rule: Opposite angles are equal.

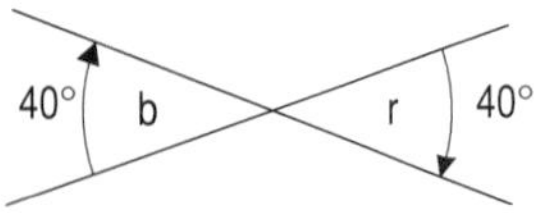

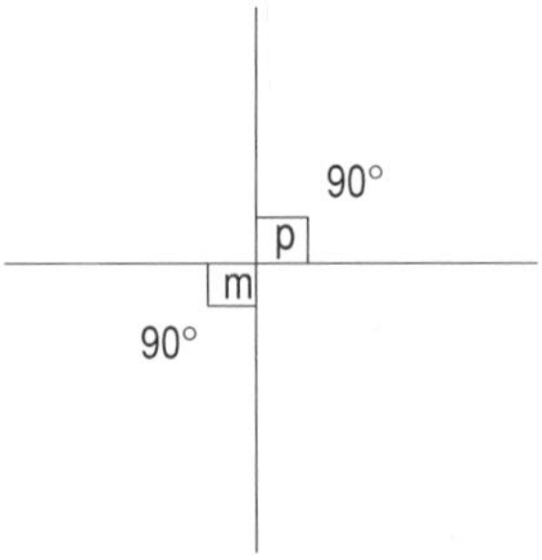

Part 2

- You can use number families to show fractions.
- The small numbers are fractions along the arrow. The big number is the fraction at the end of the arrow.

A. $\frac{2}{5}$ ——— $\frac{3}{5}$ ▶ $\frac{5}{5}$ B. $\frac{7}{9}$ ——— $\frac{2}{9}$ ▶ $\frac{9}{9}$

- For each family, you can make addition statements and subtraction statements.
- Here's an addition statement and a subtraction statement for family A:

$$\frac{2}{5} + \frac{3}{5} = \frac{5}{5}$$

$$\frac{5}{5} - \frac{3}{5} = \frac{2}{5}$$

- Fraction number families are important for working difficult word problems that have a big number of **1.**
- Here's a family with a big number of **1** and a missing small number.

$\frac{5}{9}$ ——— ■ ▶ 1

- First change **1** into a fraction. That fraction must have the same denominator as $\frac{5}{9}$.

$\frac{5}{9}$ ——— ■ ▶ $\mathbf{\frac{9}{9}}$

- Now you can find the missing small number. You work the problem $\frac{9}{9} - \frac{5}{9}$.
- The answer is $\frac{4}{9}$.

$$\frac{5}{9} \quad \frac{4}{9} \longrightarrow \frac{9}{9}$$

- Remember, the small numbers add up to **1.** All the denominators are the same.

Part 3 **For each item, make a number family. Show 1 written as a fraction with the right denominator. Figure out the missing fraction.**

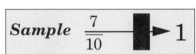

Sample $\frac{7}{10}$ ■ → 1

a. ■ $\frac{3}{12}$ → 1 b. $\frac{4}{9}$ ■ → 1 c. $\frac{2}{5}$ ■ → 1 d. ■ $\frac{6}{10}$ → 1

Part 4 **Copy the table. Figure out the missing numbers in the ratio table.**

girls	■	■
boys	3	18
children	7	■

Part 5 **Write an estimate for the answer to each problem.**

a. 320 + 706 b. 1310 + 584 c. 975 + 1180

d. 325 + 452 e. 1501 + 743 f. 319 + 2170

Part 6 **Copy each pair of problems. Multiply to figure out which answer is correct. Then complete the problem with the correct answer.**

a. $34\overline{)283}$ with 8 above; $34\overline{)283}$ with 9 above

b. $45\overline{)275}$ with 7 above; $45\overline{)275}$ with 6 above

c. $22\overline{)169}$ with 8 above; $22\overline{)169}$ with 7 above

Part 7

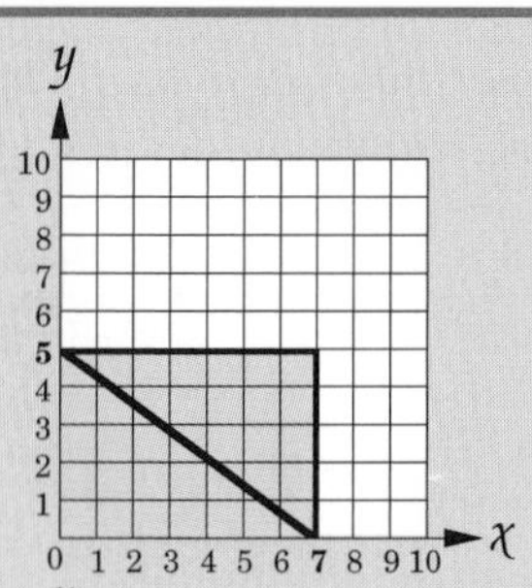

- You can make a rectangle into triangles. You just start with a rectangle and draw a DIAGONAL line. That line goes from one corner to the opposite corner.

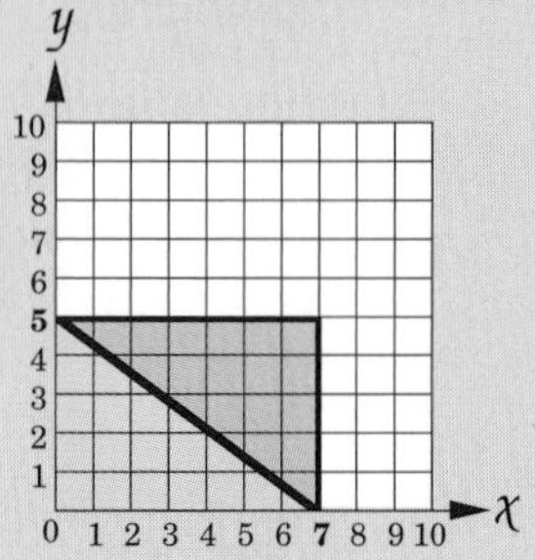

- The line divides the rectangle into two triangles. Each triangle has exactly $\frac{1}{2}$ the area of the rectangle you started with.

- Here's the equation for the area of a triangle:

$$\text{Area} \triangle = \frac{\text{base x height}}{2}$$

Part 8 Figure out the area of each shaded triangle.

Sample problem

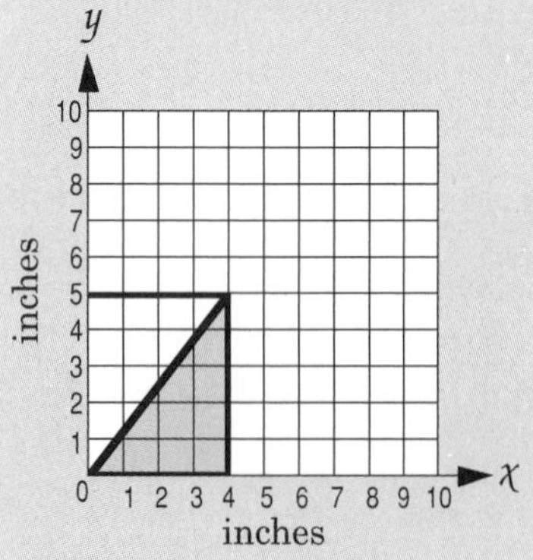

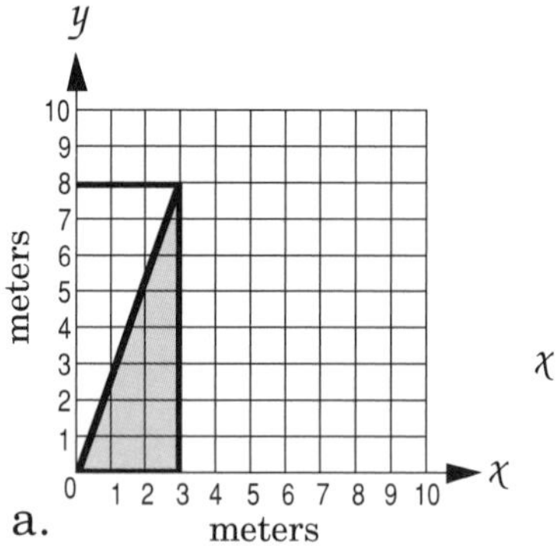

a.

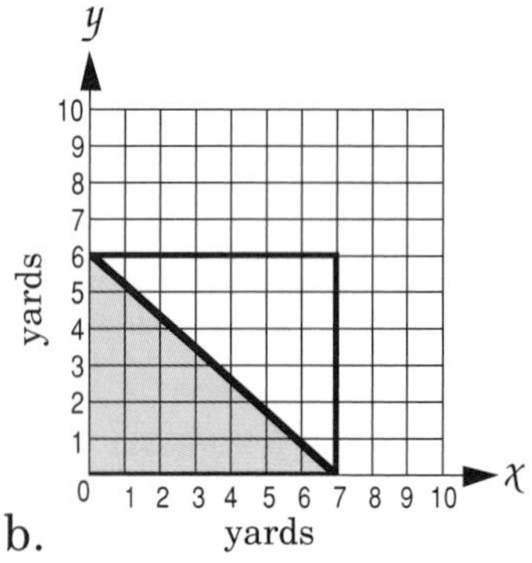

b.

Part 9

For each item, make a number family and answer the question. Remember the unit name.

a. Jane is 58 inches tall. Rose is 45 inches tall. What is the difference in the height of the two girls?

b. A truck went 584 miles. A boat went 396 miles. How much farther did the truck go than the boat went?

c. Building A was 85 feet shorter than building B. Building B was 182 feet tall. How tall was building A?

d. Load A weighed 1238 tons. Load B weighed 1575 tons. How much lighter was load A than load B?

Independent Work

Part 10

Make a table with the names and numbers the problem gives. Figure out the missing numbers. Then write answers to the items.

Women and men went to two different exercise classes. The classes were taught by Mr. Bryan and Ms. Tyler. There were 68 people enrolled in Ms. Tyler's class. There were 18 women in Mr. Bryan's class. The total number of men in both classes was 42. 44 women were in Ms. Tyler's class.

a. How many people were enrolled in Mr. Bryan's class?

b. What was the total number of men for both classes?

c. Were there more men in Mr. Bryan's class or in Ms. Tyler's class?

Part 11

For each item, write the fraction and the whole number it equals.

a. $2 = \frac{\blacksquare}{28}$ b. $\blacksquare = \frac{57}{3}$ c. $\frac{28}{7} = \blacksquare$ d. $\frac{\blacksquare}{4} = 50$ e. $\blacksquare = \frac{230}{5}$

Part 12

For each problem, write the names and a complete equation. Then answer the question.

a. The ratio of green cubes to red cubes is 5 to 4. If there are 60 red cubes, how many green cubes are there?

b. A farmer plows 2 acres every 3 hours. How long would it take the farmer to plow 60 acres?

c. For every 3 bushels of corn the farmer sells, he receives $7. If the farmer received $189 for his corn sale, how many bushels did the farmer sell?

Part 13

Trace the figure. Write the number of degrees for all the angles.

Angle t = 60°.

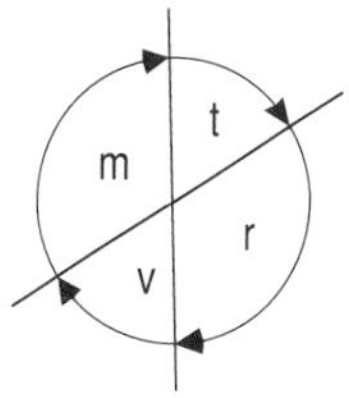

Part 14 Write an equation that shows what to multiply the first fraction by to get the second fraction. If the fractions are equal, write a simple equation below that shows the equivalent fractions.

a. $\frac{3}{8}, \frac{12}{32}$ b. $\frac{5}{4}, \frac{20}{20}$ c. $\frac{1}{9}, \frac{10}{81}$

Part 15 Rewrite each equation so it begins with the letter. Below, write what the letter equals.

a. $256 - 48 = N$

b. $\frac{315}{5} = B$

c. $8074 + 120 + 43 = J$

Part 16 For each row, write the division problem and the answer as a mixed number. Write the fraction equation for the division problem.

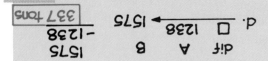

	Division	Fraction equation
a.	$7\overline{)388}$	
b.	$6\overline{)565}$	
c.	$8\overline{)300}$	

Part J

c. 85 □ → 182
dif A B
182
− 85
97 ft

d. □ 1238 → 1575
dif A B
1575
−1238
337 tons

Lesson 34

Part 1 **Figure out the area of each shaded triangle.**

a. 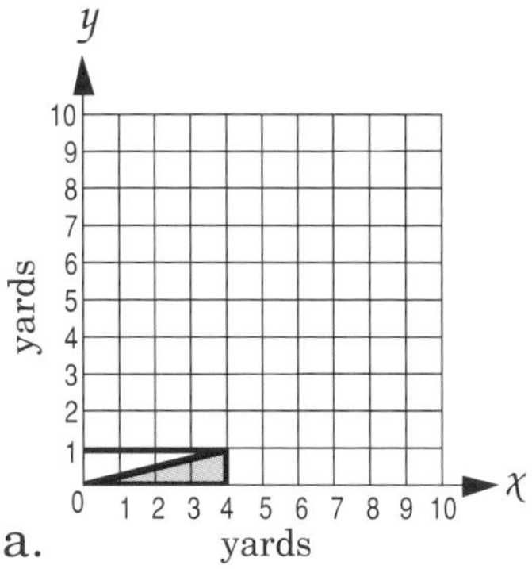

b. 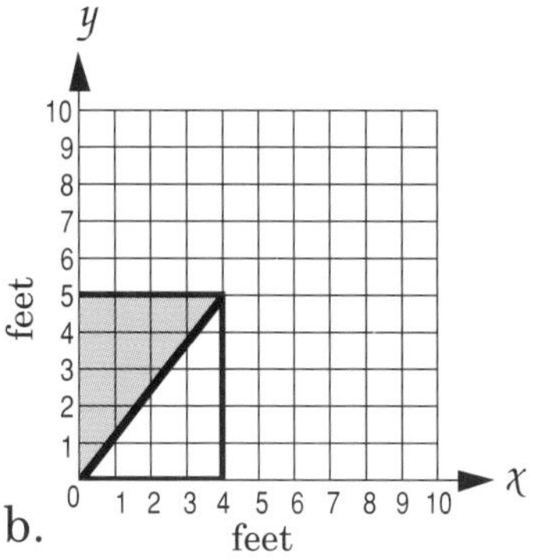

c. 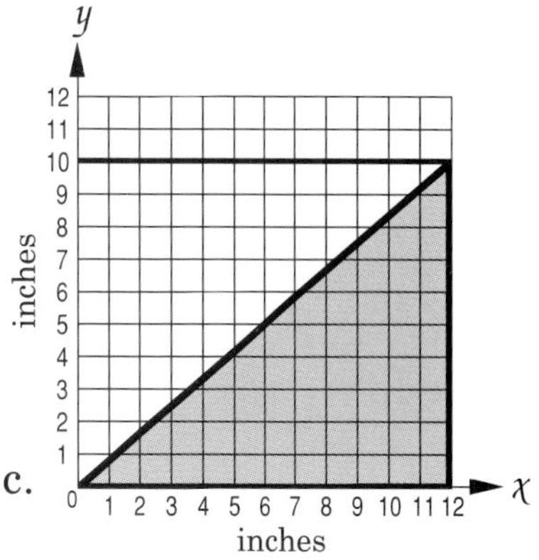

Part 2 **Copy each pair of problems. Multiply to figure out which answer is correct. Then complete the problem with the correct answer.**

a. $\begin{array}{r} 8 \\ 35\overline{)283} \end{array}$ $\begin{array}{r} 9 \\ 35\overline{)283} \end{array}$

b. $\begin{array}{r} 6 \\ 43\overline{)275} \end{array}$ $\begin{array}{r} 7 \\ 43\overline{)275} \end{array}$

Part 3 **Write an estimation answer for each problem.**

a. $\begin{array}{r} 291 \\ +\ 1296 \\ \hline \end{array}$ b. $\begin{array}{r} 1518 \\ -\ \ 699 \\ \hline \end{array}$ c. $\begin{array}{r} 2560 \\ -\ \ 419 \\ \hline \end{array}$ d. $\begin{array}{r} 1829 \\ +\ 1055 \\ \hline \end{array}$

Part 4 **Copy each pair of problems. Multiply to figure out which answer is correct. If the remainder is too big, circle it. Then complete the problem with the correct answer.**

Sample

$\begin{array}{r} 6 \\ 34\overline{)241} \\ -\ 204 \\ \hline (37) \end{array}$ $\begin{array}{r} 7\frac{3}{34} \\ 34\overline{)241} \\ -\ 238 \\ \hline 3 \end{array}$

a. $\begin{array}{r} 5 \\ 61\overline{)348} \end{array}$ $\begin{array}{r} 4 \\ 61\overline{)348} \end{array}$

b. $\begin{array}{r} 6 \\ 47\overline{)367} \end{array}$ $\begin{array}{r} 7 \\ 47\overline{)367} \end{array}$

Part 5 **For each item, write a number family. Show 1 written as a fraction with the right denominator. Figure out the missing fraction.**

a. $\frac{6}{20}$ ■ → 1 b. $\frac{11}{12}$ ■ → 1 c. ■ $\frac{5}{28}$ → 1 d. $\frac{7}{35}$ ■ → 1

Part 6 **Copy each table. Figure out the missing numbers in each ratio table.**

a.

men	6	■
women	1	■
people	■	56

b.

men	■	■
women	3	27
people	5	■

c.

men	■	42
women	3	■
people	9	■

Part 7 **Find the area and perimeter of each parallelogram.**

a.

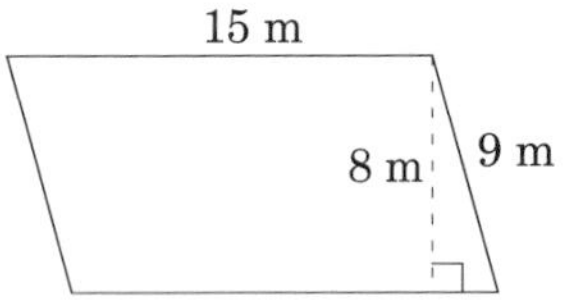

b.

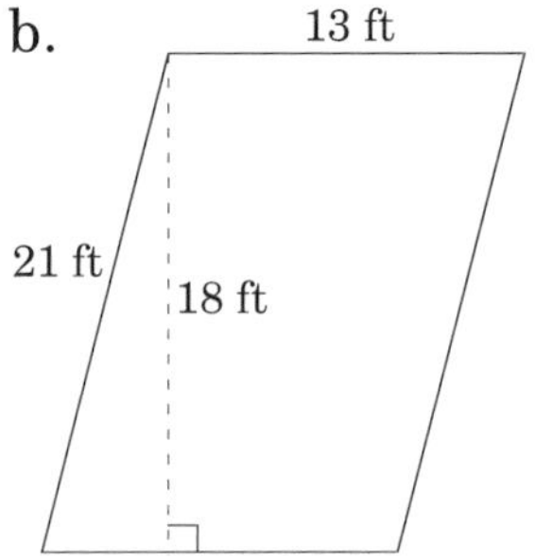

Independent Work

Part 8 **For each item, make a number family. Answer the question the problem asks.**

a. The bull weighs 2130 pounds. The calf weighs 502 pounds. How much lighter is the calf than the bull?

b. A car traveled 234 miles. A truck traveled 186 miles farther than the car traveled. How far did the truck travel?

c. The snow on Mt. Hood was 111 inches deeper in February than in July. In July, the snow was 13 inches deep. What was the depth of the snow in February?

Part 9 **Trace the figure. Write the number of degrees for each angle.**

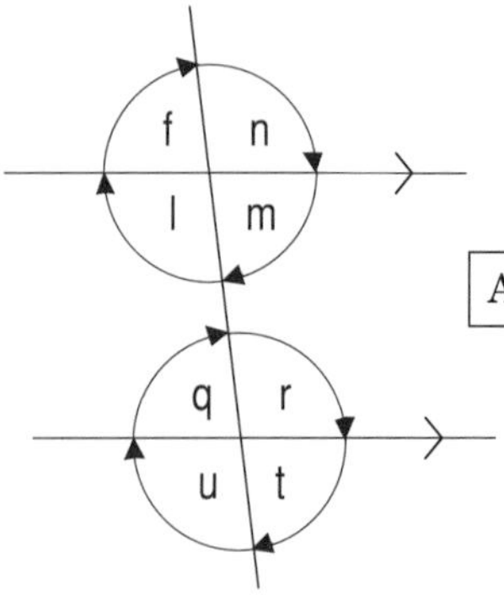

Angle n = 97°.

Part 10 For each item, make a number family. Answer the question the problem asks.

a. The crew washed 27 cars and 89 vehicles that were not cars. How many vehicles did the crew wash in all?

b. 500 students attended Emerson School. 338 of those students lived more than a mile from school. How many students did not live more than a mile from school?

c. In the morning, a train traveled 161 miles. During the afternoon, the train traveled even farther. By the end of the afternoon, the train had traveled 402 miles. How many miles did the train travel during the afternoon?

Part 11 Find the area and the perimeter of each rectangle.

a. 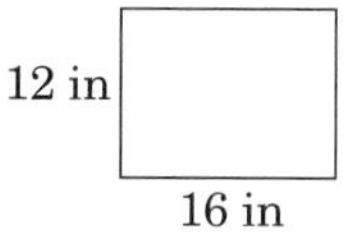

b. 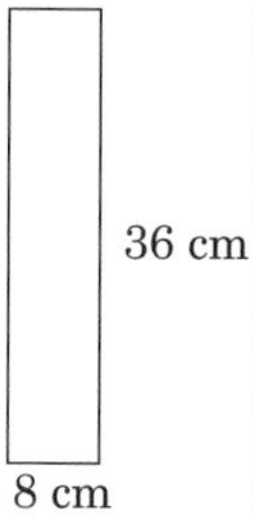

c. 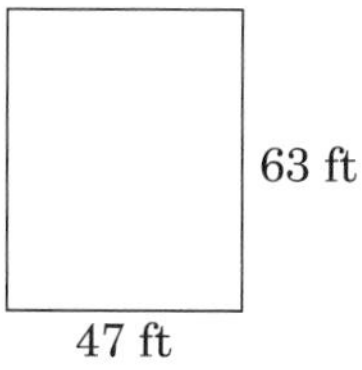

Part 12 For each item, make a table and complete it.

a. Add 5 to each number. Then multiply the sum by 2.
0, 6, 10

b. Subtract each number from 20. Then multiply the difference by 2.
19, 5, 12

c. Multiply each number by 3. Then subtract 1.
50, 2, 7

Part 13 Work each problem. Write each answer with a unit name.

a. Each softball costs $4. John bought 12 softballs. How much did he pay?

b. A parcel of land was divided into 6 parts that were the same size. The original parcel was 360 acres. How many acres were in each of the 6 parts?

c. A barrel of fish weighed 780 pounds. The fish were arranged in piles that each weighed 6 pounds. How many piles were there?

d. Each pillow weighed 48 ounces. There were 4 pillows in a carton. What was the total weight inside the carton?

Part 14 Copy and complete each equation.

a. $\frac{5}{6} \times \frac{7}{6} = \blacksquare$

b. $\frac{3}{4}(\blacksquare) = \frac{21}{8}$

c. $\frac{13}{4} - \frac{8}{4} - \frac{5}{4} = \blacksquare$

d. $\frac{\blacksquare}{3} = 6$

e. $\frac{2}{5} = \frac{\blacksquare}{200}$

f. $\blacksquare = \frac{42}{3}$

g. $\frac{5}{3} + \frac{1}{3} - \frac{2}{3} = \blacksquare$

Part J

b. $A\Delta = \frac{b \times h}{2}$

$A\Delta = \frac{4 \times 5}{2} = \frac{20}{2}$

$A\Delta = 10$ sq ft

c. $A\Delta = \frac{b \times h}{2}$

$A\Delta = \frac{12 \times 10}{2} = \frac{120}{2}$

$A\Delta = 60$ sq in

Lesson 35

Part 1 **Copy each table. Write the names where they go. Then complete the tables.**

a. The names are **children, girls** and **boys.** The number for boys in the first column is 4.

	4	68
	■	■
	12	■

b. The names are **wet shirts, shirts** and **dry shirts.** The number of wet shirts in the first column is 7.

	■	24
	7	■
	11	■

c. The names are **trees, short trees** and **tall trees.** The number of short trees in the first column is 3.

	4	■
	3	■
	■	42

Part 2

- The correct answer to a division problem may have a remainder.
- If the answer above the division sign is **too big,** the number you get when you multiply is too big. You can't subtract.
- Here's a problem with an answer that is **too big:**

$$\begin{array}{r} 7 \\ 25\overline{)163} \\ \underline{-175} \end{array}$$

- If the answer above the division sign is **too small,** the number you get when you multiply is too small. So the remainder is bigger than the number you're dividing by.
- Here's the problem with an answer that is **too small:**

$$\begin{array}{r} 5 \\ 25\overline{)163} \\ \underline{-125} \\ 38 \end{array}$$

- If the answer above the division sign is **correct,** you **can subtract** and the remainder is smaller than the number you're dividing by Here's the problem with the **correct** answer:

$$\begin{array}{r} 6\frac{13}{25} \\ 25\overline{)163} \\ \underline{-150} \\ 13 \end{array}$$

Part 3

Figure out if each answer is too big or too small. Then work the problem with the correct answer.

a. $\begin{array}{r} 5 \\ 18\overline{)82} \end{array}$ b. $\begin{array}{r} 3 \\ 33\overline{)145} \end{array}$ c. $\begin{array}{r} 7 \\ 31\overline{)188} \end{array}$ d. $\begin{array}{r} 6 \\ 59\overline{)417} \end{array}$

Part 4

Make a fraction number family for each word problem. Answer the question the problem asks.

> ***Sample*** $\frac{2}{5}$ of the children in a class are boys. What fraction of the children are girls?

a. $\frac{1}{9}$ of the trees in a forest are pines. What's the fraction for the trees that are not pines?

b. In Sage City, $\frac{8}{10}$ of the days are dry. What's the fraction for days that are rainy?

Part 5

Use an estimate to check the answer to each problem. If the answer is wrong, copy the problem and rework it.

a. $\begin{array}{r} 1492 \\ -\ 655 \\ \hline 837 \end{array}$ b. $\begin{array}{r} 2002 \\ -\ 878 \\ \hline 2780 \end{array}$ c. $\begin{array}{r} 478 \\ +\ 910 \\ \hline 1388 \end{array}$ d. $\begin{array}{r} 1441 \\ -\ 795 \\ \hline 846 \end{array}$

e. $\begin{array}{r} 423 \\ +\ 1965 \\ \hline 2388 \end{array}$ f. $\begin{array}{r} 1595 \\ -\ 284 \\ \hline 1311 \end{array}$ g. $\begin{array}{r} 4762 \\ +\ 1943 \\ \hline 3809 \end{array}$ h. $\begin{array}{r} 2119 \\ +\ 4391 \\ \hline 6510 \end{array}$

Part 6

Find the area of each figure.

a.
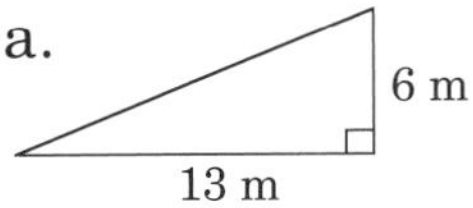

b.
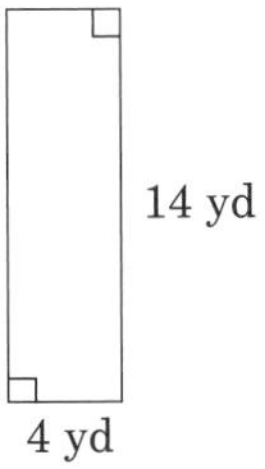

c.
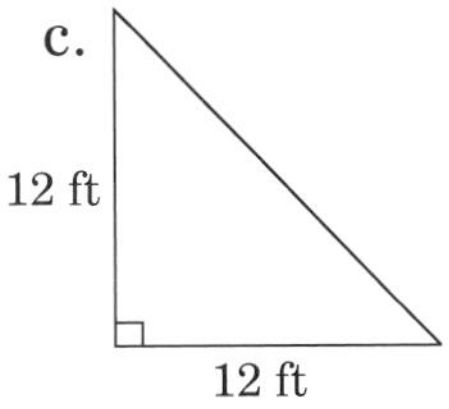

d.
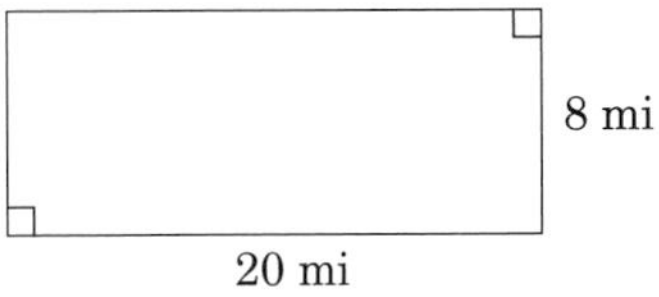

Independent Work

Part 7 **Make a table with the names and numbers the problem gives. Figure out the missing numbers. Then write answers to the items.**

There were children and adults at two different amusement centers. The centers were Joyland and Big Ride. There were 81 adults at Joyland and 162 adults at Big Ride. There were 20 children at Joyland and 40 children at Big Ride.

a. How many people were at Big Ride?

b. What was the total number of adults for both centers?

c. Were there more people at Joyland or Big Ride?

Part 8 **Answer each question.**

a. 827 men and women worked in a factory. 584 of the workers were women. How many of the workers were men?

b.

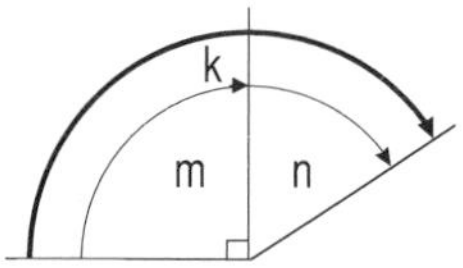

Angle n is 57°.
What is angle k?

Part 9 **For each item, make a number family. Answer the question the problem asks.**

a. In a library there are 847 fiction books. The library contains 1403 nonfiction books. How many more nonfiction books are there than fiction books?

b. A snake was 158 millimeters longer than a worm. The worm was 200 millimeters long. How long was the snake?

Part 10 **Trace the figure. Write the number of degrees for each angle.**

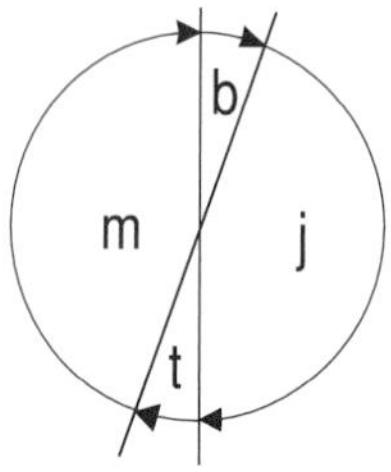

Angle j is 160°.

Part 11 **For each item, write the names and complete equation. Then answer the question.**

a. The ratio of collars to cats was 2 to 9. If there were 44 collars, how many cats were there?

b. A machine made 17 buttonholes every 5 seconds. How many buttonholes would the machine make in 60 seconds?

c. An oven could bake 6 loaves of bread every 24 minutes. How many minutes would it take the oven to bake 180 loaves of bread?

Part 12 **Rewrite each equation so it begins with the letter. Below, write what the letter equals.**

a. 684 + 7 – 250 = T

b. 132 x 51 = R

c. 80,490 – 9,180 = M

Part 13 **Copy and complete the table. Write the division problems with the answer as a whole number or a mixed number.**

	Fraction equation	Division
a.	$\frac{286}{5}$ = ■	
b.		$9\overline{)573}$
c.		$4\overline{)478}$
d.	$\frac{560}{10}$ = ■	

Lesson 36

Part 1 **Copy each table. Write the names where they go.**

a. The names are apples **with worms, apples** and apples **without worms.** In the first column, the number for apples with worms is 6.

	6	■
	■	■
	9	36

b. The names are **shoes, clean** shoes and **dirty** shoes. In the first column, the number for clean shoes is 2.

	■	144
	8	■
	10	■

c. The names are **boys, children** and **girls.** The number for girls in the first column is 5.

	5	■
	7	■
	■	120

d. The names are **closed** doors, **open** doors and **doors.** The number for open doors in the first column is 1.

	3	276
	■	■
	4	■

e. The names are **women, short** women and **tall** women. The number for tall women in the first column is 5.

	5	■
	■	44
	9	■

Part 2 **Use an estimate to check the answer to each problem. If the answer is wrong, copy the problem and rework it.**

a. $\begin{array}{r} 3587 \\ -\ \ 428 \\ \hline 2661 \end{array}$ b. $\begin{array}{r} 1124 \\ -\ \ 695 \\ \hline 429 \end{array}$ c. $\begin{array}{r} 910 \\ +\ 1874 \\ \hline 2784 \end{array}$ d. $\begin{array}{r} 1290 \\ -\ \ 405 \\ \hline 1695 \end{array}$

e. $\begin{array}{r} 1418 \\ +\ 2475 \\ \hline 4683 \end{array}$ f. $\begin{array}{r} 1959 \\ +\ \ 695 \\ \hline 2654 \end{array}$ g. $\begin{array}{r} 526 \\ -\ 278 \\ \hline 248 \end{array}$ h. $\begin{array}{r} 5980 \\ +\ 2419 \\ \hline 8399 \end{array}$

Part 3 **For each item, make a number family.**

a. 25 of the shirts in a department store were blue. The rest were not blue. The store had a total of 300 shirts. How many were not blue?

b. There were 25 fewer tan shirts than white shirts. There were 30 tan shirts. How many white shirts were there?

c. A bull weighed 2030 pounds. A cow weighed 1658. What was the difference in their weight?

d. Frank bought materials. He spent $45 on paint. He spent $21 on brushes. What was the total amount he spent?

e. The bull weighed 307 pounds more than the cow. The cow weighed 789 pounds. How much did the bull weigh?

f. A store carried hard-cover books and soft-cover books. The store had 8003 books. 808 were hard-cover. How many soft-cover books were in the store?

Part 4 **Find the area of each figure.**

a.

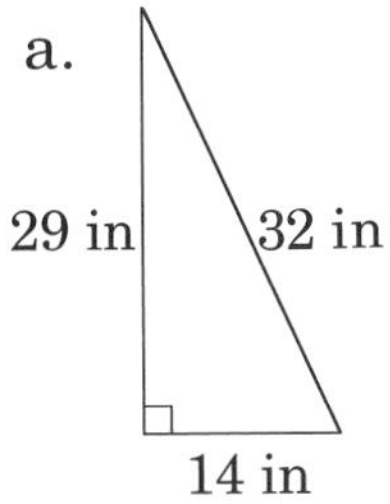

b.

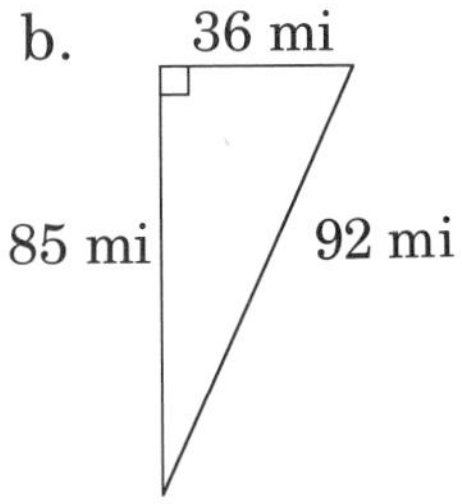

c.

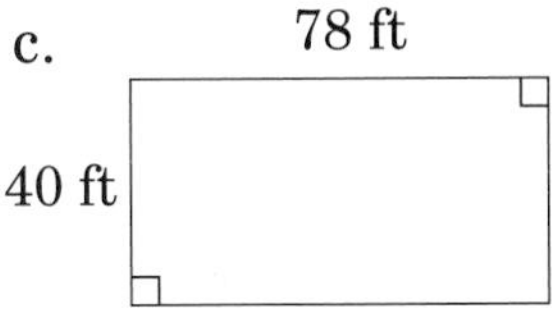

Part 5 **Make a fraction number family for each word problem. Box the fraction that answers the question the problem asks.**

a. $\frac{1}{5}$ of the vehicles are not cars. What fraction of the vehicles are cars?

b. $\frac{2}{7}$ of the fish are perch. What fraction of the fish are not perch?

c. $\frac{3}{10}$ of the cups are not cracked. What fraction of the cups are cracked?

Part 6

Figure out if each answer is too big or too small. Then work the problem with the correct answer.

a. $64\overline{)345}$ with 6 written above

b. $37\overline{)227}$ with 5 written above

c. $42\overline{)240}$ with 4 written above

Independent Work

Complete the tables in part 1.

Part 7

Find the area and perimeter of each figure.

a.

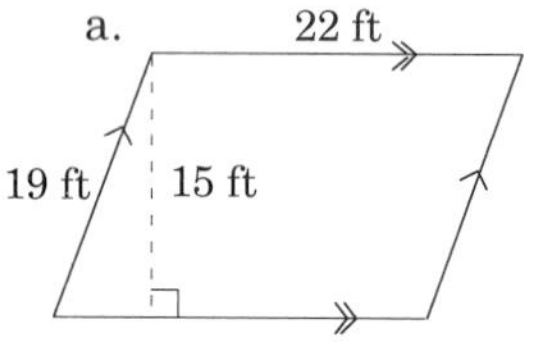

b.

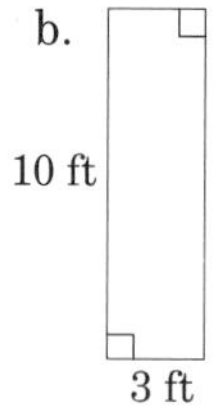

Part 8

For each item, write the names and a complete equation. Then answer the questions.

a. On a tree, there are 4 green leaves for every 3 brown leaves. There are 372 green leaves. How many brown leaves are on the tree?

b. A baseball travels 80 feet every 2 seconds. How far will the ball travel in 6 seconds?

c. John read 7 pages for every 2 pages his younger sister read. If John read 483 pages, how many pages did his sister read?

Part 9

For each item, write the fraction and the whole number it equals.

a. $\frac{246}{6} = \blacksquare$ b. $\frac{732}{5} = \blacksquare$

c. $\frac{\blacksquare}{3} = 132$

Part 10

Copy and complete the table.

	Division	Multiplication	Fraction equation
a.	$7\overline{)126}$		
b.	$8\overline{)224}$		
c.	$9\overline{)342}$		

Part 11

Make the complete number family with three fractions.

a. ——— $\frac{3}{10}$ → 1 b. $\frac{4}{5}$ ——— → 1 c. ——— $\frac{3}{23}$ → 1 d. $\frac{2}{9}$ ——— → 1

Part 12

Complete each equation. Below, write a simple equation for the equivalent fractions.

a. $\frac{3}{8} = \frac{36}{\blacksquare}$ b. $\frac{7}{1} = \frac{\blacksquare}{30}$ c. $\frac{12}{5} = \frac{36}{\blacksquare}$ d. $\frac{2}{9} = \frac{22}{\blacksquare}$

Part J

c. dif / cow / bull: □, 1658 → 2030

d. paint / brushes / total: 45, 21 → □

e. dif / cow / bull: 307, 789 → □

f. hard-cover / soft-cover / books: 808, □ → 8003

Part K

a. $A\triangle = \frac{b \times h}{2}$

$A\triangle = \frac{14 \times 29}{2} = \frac{406}{2}$

$A\triangle = 203$ sq in

b. $A\triangle = \frac{b \times h}{2}$

$A\triangle = \frac{36 \times 85}{2} = \frac{3060}{2}$

$A\triangle = 1530$ sq mi

c. $A\square = b \times h$

$A\square = 78 \times 40$

$A\square = 3120$ sq ft

Part L

a. 64)345 — 6, −384 (crossed out); 64)345 — 5, −320, 25 → $5\frac{25}{64}$

b. 37)227 — 5, −185, (42) (crossed out); 37)227 — 6, −222, 5 → $6\frac{5}{37}$

c. 42)240 — 4, −168, (72) (crossed out); 42)240 — 5, −210, 30 → $5\frac{30}{42}$

I don't know what the teacher means about not being able to subtract. I can subtract for every problem.
9 20/52
52)448
- 1468
20

Lesson 37

Part 1 **Use an estimate to check the answer to each problem. If the answer is wrong, copy the problem and rework it.**

a. 47 − 18 = 29	b. 91 − 47 = 138	c. 6775 + 438 = 6313	d. 49 + 57 = 106
e. 394 + 808 = 1202	f. 70 − 27 = 43	g. 99 + 70 = 29	h. 1104 − 410 = 690

Part 2

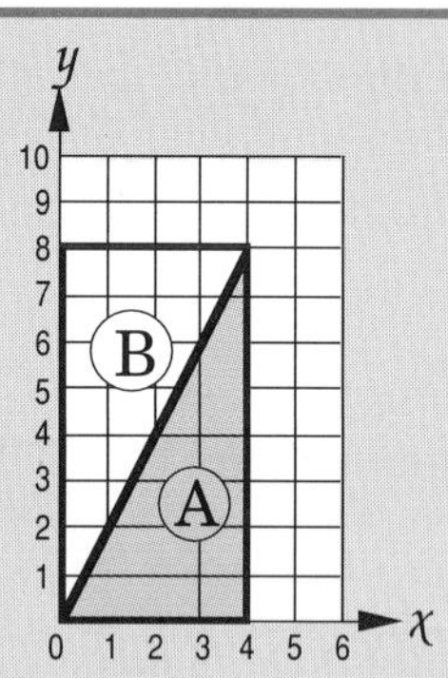

- You've found the area of triangles that have a 90° angle.
- You've used the equation: $\text{Area} \triangle = \dfrac{\text{base x height}}{2}$
- Triangle A and triangle B are identical. Each is half the area of the rectangle with a base of 4 units and a height of 8 units.
- You can use the same equation to show that the area of **any triangle** is half the area of a **parallelogram** with the same base and same height.

- Triangle C has no 90° angle.
- We can make a **parallelogram** by combining triangle C with another triangle that is exactly the same size and shape.

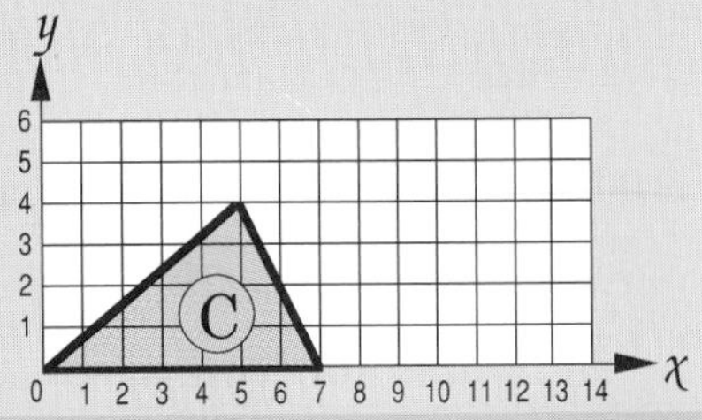

- Triangle C is combined with triangle D to form a parallelogram.
- The base is 7 and the height is 4. So the whole parallelogram has an area of 28 square units.
- The area of triangle C is half that amount–14 square units.

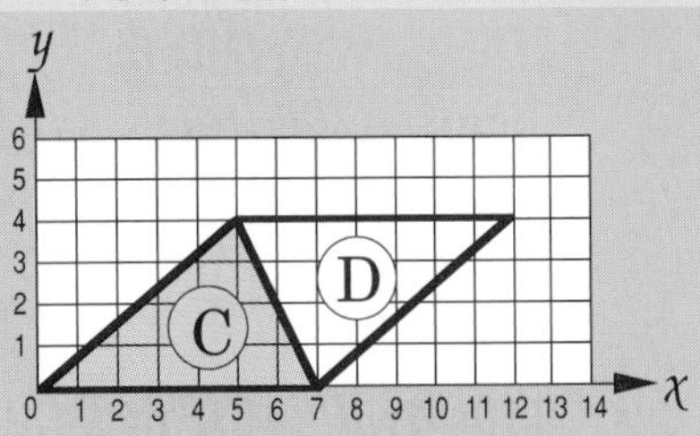

- Remember, the area of any triangle is half the area of a parallelogram with the same base and the same height.

Part 3 **Figure out the area of each shaded triangle.**

a.

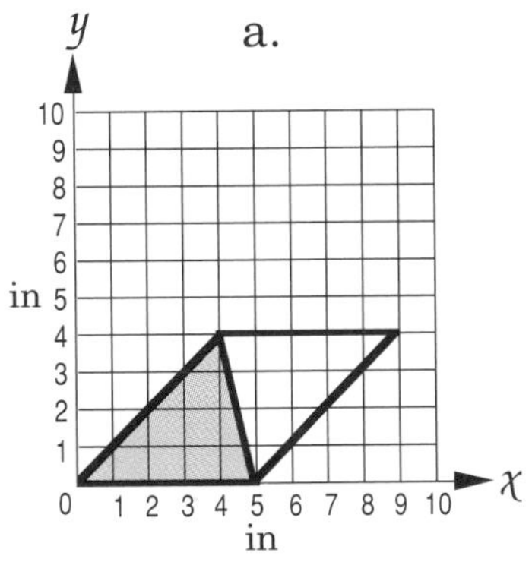

b.

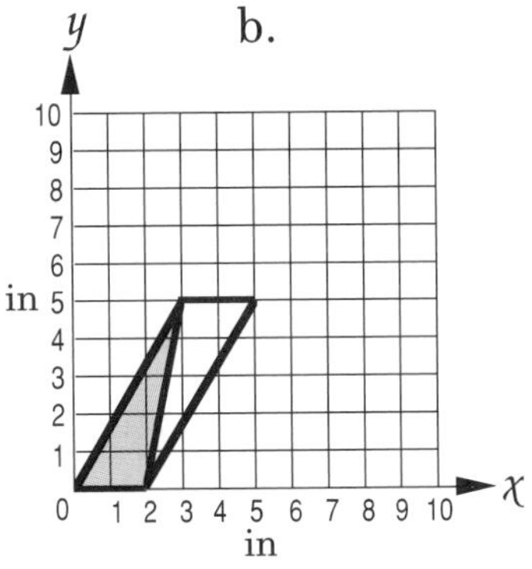

c.

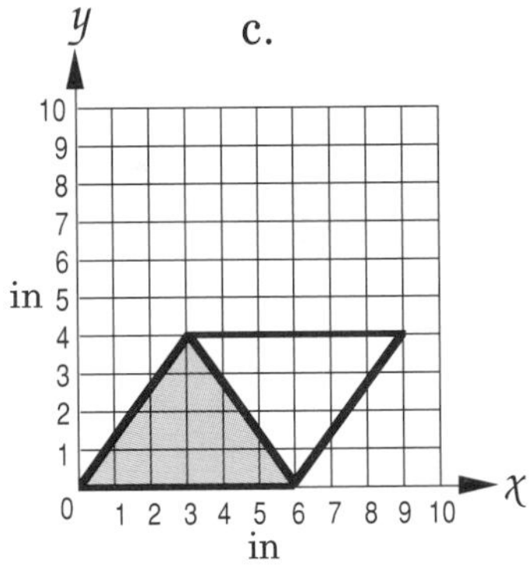

Part 4 **Use an estimate to work each division problem.**

a. $34\overline{)175}$ b. $92\overline{)278}$ c. $43\overline{)271}$

Part 5

- You've worked word problems that you solve by making ratio equations. The problems you've worked have **two names.**
- You can work problems that have **three names.** To work those problems, you first make a ratio table that shows the three names.
- Here's a problem:

The ratio of rainy days to dry days is 1 to 8. There was a total of 54 days. How many days were rainy? How many days were dry?

	ratio	
rainy		
dry		
days		

- The first sentence tells the ratio numbers for rainy days and dry days. Those numbers go in the first column.
- The problem tells that there was a total of 54 days. That number goes in the second column.

	ratio	
rainy	1	
dry	8	
days		54

- You can figure out the total in the first column and the missing numbers in the second column.
- Here's the table with no missing numbers:

	ratio	
rainy	1	6
dry	8	48
days	9	54

- Remember, the ratio numbers go in the first column.

Part 6

Make a ratio table. Answer the questions for each item.

a. There were **perch** and **bass** in a pond. For every 10 **fish,** 7 were perch. There were 900 bass in the pond.

1. How many perch were in the pond?
2. How many total fish were in the pond?

a.

	ratio	

b. There were **perch** and **bass** in a pond. The ratio of perch to bass was 4 to 5. There was a total of 180 **fish** in the pond.

1. How many bass were in the pond?
2. How many perch were in the pond?

c. There were **parts** in a junkyard. For every 8 parts, 7 were **rusty.** There were 60 parts that were **not rusty.**

1. How many parts were rusty?
2. How many total parts were in the yard?

Part 7

Make a fraction number family for each word problem. Box the fraction that answers the question.

a. $\frac{2}{9}$ of the trees are pines. What fraction of the trees are not pines?

b. $\frac{5}{8}$ of the adults in the factory were women. What fraction of the adults were men?

c. $\frac{1}{7}$ of the food the horse ate was not grass. What's the fraction of food that was grass?

d. Jane scored $\frac{3}{5}$ of the team's points. What fraction of the points did her teammates score?

Independent Work

Part 8

Figure out if each answer is too big or too small. Then work the problem with the correct answer.

a. $\begin{array}{r}4\\74\overline{)257}\end{array}$ b. $\begin{array}{r}7\\81\overline{)558}\end{array}$ c. $\begin{array}{r}5\\57\overline{)351}\end{array}$

Part 9

Find the area of the parallelogram and triangles.

a.

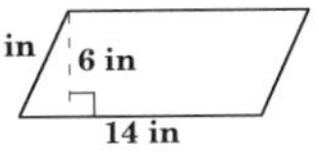

b.

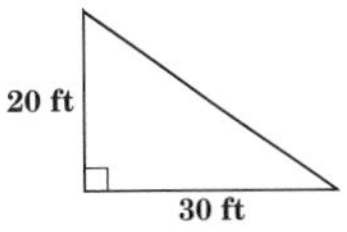

c.

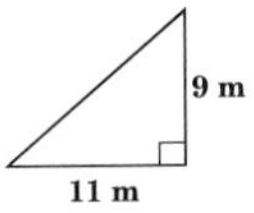

Part 10

Copy and complete each table.

a. The names for the table are **sharpened** pencils, **pencils** and **unsharpened** pencils. The number for unsharpened pencils in the first column is 3.

	3	■
	■	32
	7	■

b. The names for the table are **good** tires, **flat** tires, and **tires.** The number for good tires in the first column is 11.

	11	■
	1	■
	■	144

Part 11

Trace each figure. Show the degrees for each lettered angle.

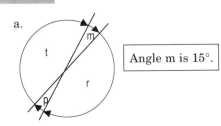

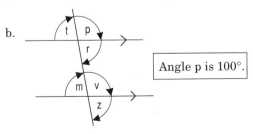

a.

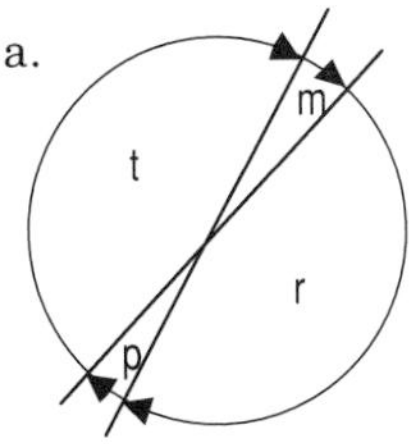

Angle m is 15°.

b.

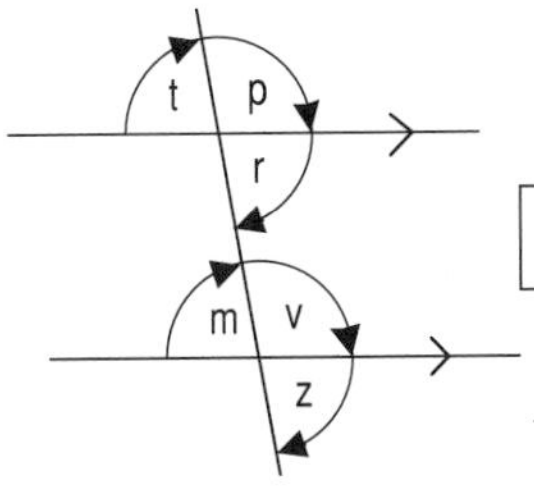

Angle p is 100°.

Part 12 **Answer each question.**

a. A bull weighs 2174 pounds. A cow weighs 1423 pounds. How much lighter is the cow than the bull?

b. A bull and a cow are in the back of a truck. The bull weighs 2174 pounds and the cow weighs 1423 pounds. What is the total weight of the animals in the back of the truck?

c. In the afternoon, the train traveled 124 miles farther than it traveled in the morning. The train traveled 213 miles in the afternoon. How far did it travel in the morning?

d. In a garage, there were cars that needed repairs and cars that were repaired. 31 cars needed repairs. The total number of cars in the shop was 40. How many cars did not need repairs?

Lesson 38

Part 1

- You've worked problems that divide by two-digit numbers. You can estimate the answer by covering the last digit of each value in the problem.
- Here's 478 divided by 79: $79\overline{)478}$
- If you cover the last digit of each number, you get the estimation problem, 47 divided by 7. $7\overline{)47}$
- You can get a more accurate estimation problem if you **round the number you divide by** to the **nearest ten.**
- Here's the division problem based on the rounding: $8\overline{)47}$

- Here's a division problem: $49\overline{)308}$
- You're dividing by 49. The estimated answer is 6. $\begin{array}{r}6\\5\overline{)30}\end{array}$ $\begin{array}{r}6\\49\overline{)308}\end{array}$

- Here's a different problem: $41\overline{)308}$
- You're dividing by 41. The estimated answer is 7. $\begin{array}{r}7\\4\overline{)30}\end{array}$ $\begin{array}{r}7\\41\overline{)308}\end{array}$

Part 2

Use an estimate to work each problem.

a. $47\overline{)208}$ b. $29\overline{)241}$ c. $52\overline{)368}$

Part 3 **Make a number family for each word problem. Answer the question the problem asks.**

a. In a garden, there were flying insects and insects that could not fly. There was a total of 560 insects. If 211 were flying insects, how many were insects that could not fly?

b. In a park, there were 118 fewer flying insects than insects that could not fly. There were 871 insects that could not fly. How many flying insects were there?

c. In a park, there were red ants and black ants. There were 456 red ants and 119 black ants. How many more red ants were there than black ants?

d. In a yard, there were red ants and black ants. There were 581 red ants and 222 ants that were not red. How many ants were there in all?

Part 4 **For each item, write an estimation problem and the answer.**

4686 rounds to 5000
4329 rounds to 4000
379 rounds to [redacted]
608 rounds to [redacted]

a.	b.	c.	d.
680	28	5600	240
5290	508	176	266
+ 390	+ 1299	+ 2689	+ 8919

Part 5 **Make a ratio table. Answer the questions for each item. Write your answers as a number and a unit name.**

a. There are **boys** and **girls** in a school. There are 3 girls for every 5 **children.** There are 402 boys in the school.
   1. How many children are in the school?
   2. How many girls are in the school?

b. There are **blue** balloons and **red** balloons at a party. The ratio of blue balloons to red balloons is 5 to 3. There is a total of 240 **balloons.**
   1. How many blue balloons are there?
   2. How many red balloons are there?

Part 6 Make a fraction number family for each item.

Sample

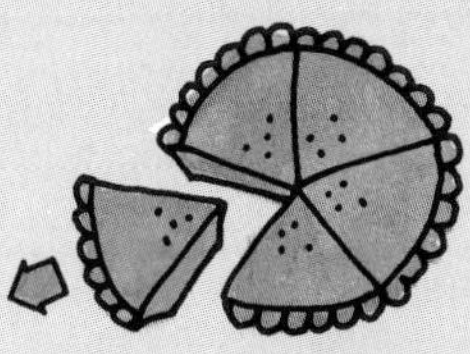

removed $\frac{1}{5}$ → remaining $\frac{4}{5}$ → pie $\frac{5}{5}$

a.

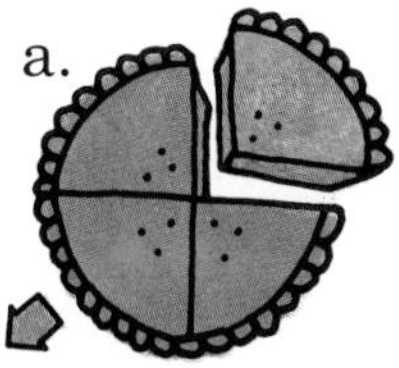

b.

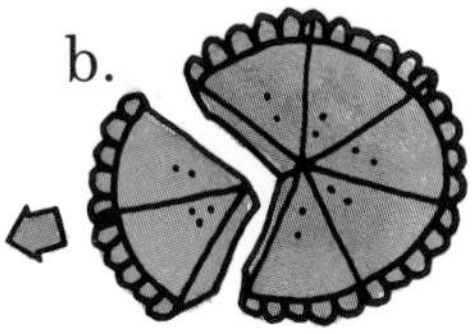

c.

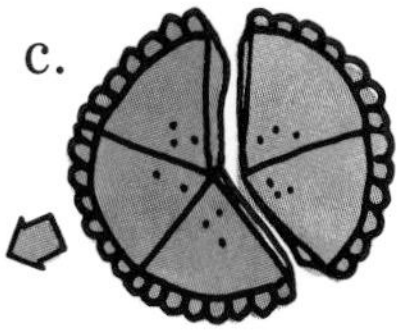

d.

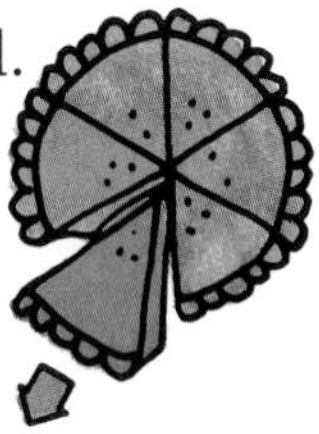

Part 7 Find the area of each triangle.

a.

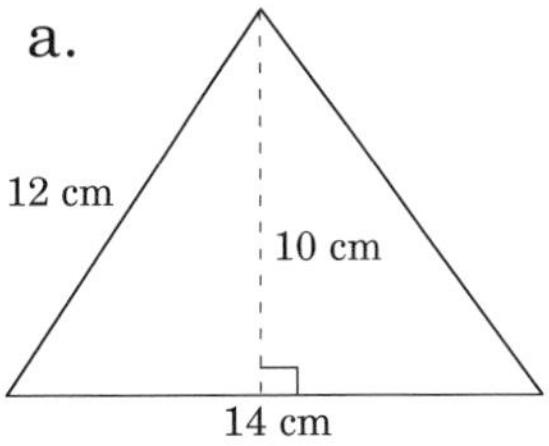

b.

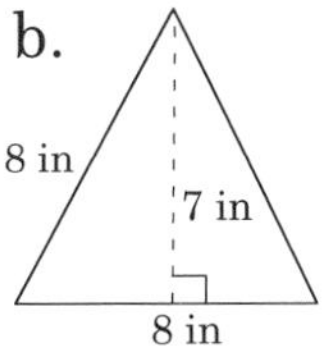

c.

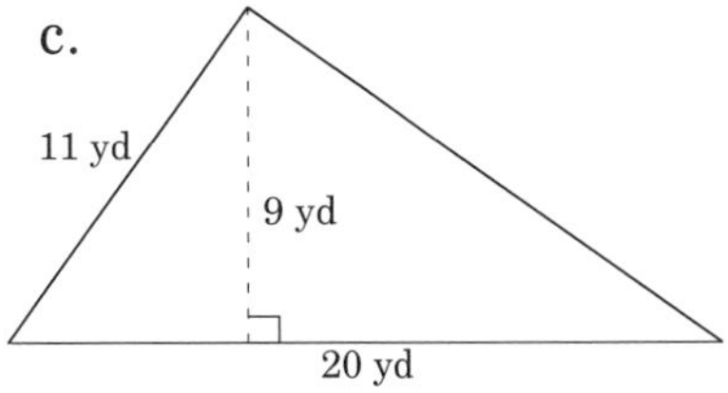

Independent Work

Part 8 Find the perimeter of the parallelogram and triangles.

a.

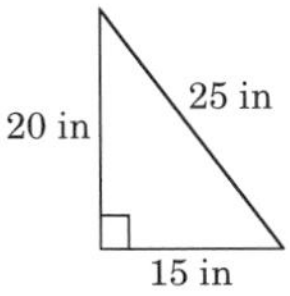

b.

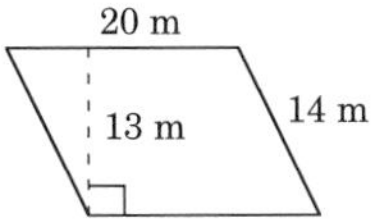

c.

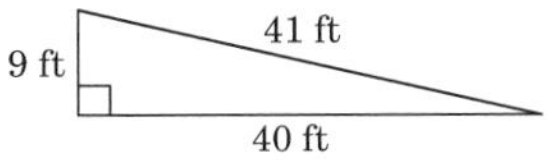

Part 9 For each item, make a number family. Answer the question.

a. Mr. Smith's farm had 250 more acres than Mr. Thompson's farm. Mr. Smith's farm had 940 acres. How many acres were in Mr. Thompson's farm?

b. On Tuesday, it rained at 47 of the cities that were surveyed. It did not rain at 82 of the cities that were surveyed. How many cities were surveyed?

c. The yellow car cost $11,200. The red car cost $15,300. What was the difference in their cost?

d. In a field, 28 watermelons were not ripe. The rest were ripe. There were 348 watermelons in the field. How many were ripe?

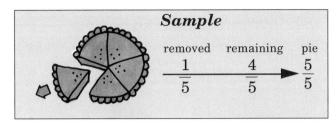

Part 10 **Copy and work each problem. If the answer is too big or too small, work the problem again with the correct answer.**

a. $68\overline{)358}$ with 4 written above

b. $92\overline{)358}$ with 3 written above

c. $41\overline{)358}$ with 9 written above

Part 11 **Find the degrees for each lettered angle.**

a.

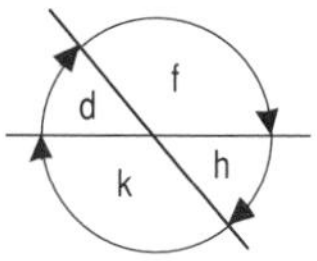

Angle h is 50°.

b.

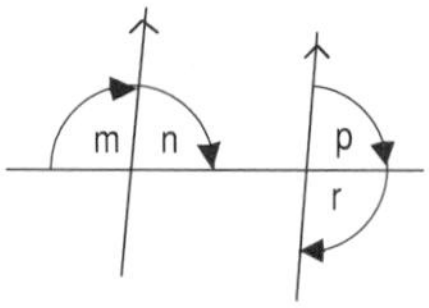

Angle p is 85°.

Part 12 **Rewrite and complete each number family.**

a. $\frac{6}{17}$ ——▶ 1

b. ———— $\frac{3}{5}$ ▶ 1

c. $\frac{4}{20}$ ——▶ 1

Part 13 **Copy and complete each table.**

a. The names for the table are **broken** cups, **cups** and cups that are **not broken.** The number for broken cups in the first column is 8.

	■	■
	3	■
	■	77

b. The names for the table are **children, girls** and **boys.** The number for girls in the first column is 3.

	■	90
	4	■
	■	■

Part 14 **For each item, make a table and complete it.**

a. Multiply the number by 21.
Subtract the product from 299.
13, 0, 10

b. Add 32 to each number.
Divide the sum by 5.
8, 23, 3

Lesson 39

Part 1 Find the area and perimeter of each figure.

a.

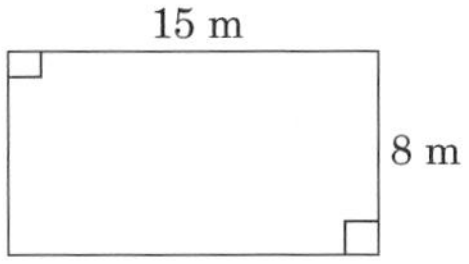

b.

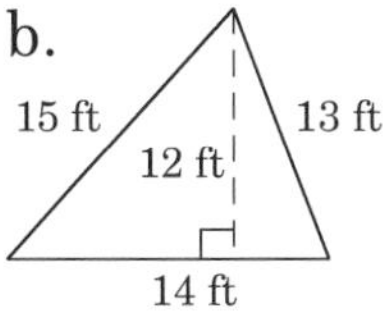

c.

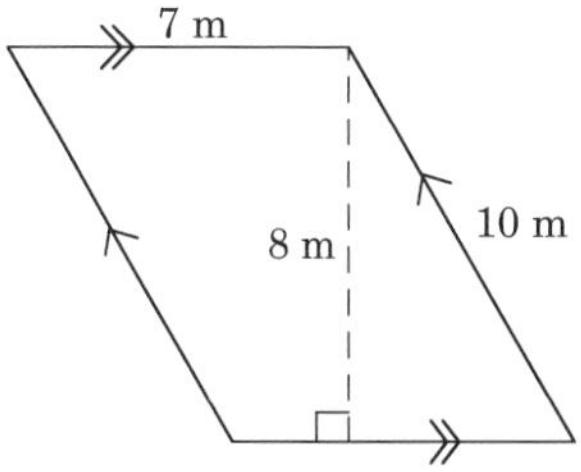

Part 2 For each item, make a number family. Answer the question.

a. Hill Park had 340 flowers. Dill Park had 68 flowers. How many fewer flowers were in Dill Park than in Hill Park?

b. Hill Park and Mill Park had flowers. The total number of flowers for both parks was 703. If Mill Park had 400 flowers, how many flowers were in Hill Park?

c. Gill Park and Dill Park had trees. There were 567 more trees in Gill Park than in Dill Park. If Gill Park had 893 trees, how many trees were in Dill Park?

d. There were large trees and small trees in Hill Park. 87 were large. If there were 203 trees in Hill Park, how many were small?

Part 3 Use an estimate to work each problem.

a. $19\overline{)94}$ b. $82\overline{)501}$ c. $68\overline{)499}$

Part 4

- When you round the first digit of a numeral, the rounded value may be more than 9. When it is, you write the rounded value as 10.
- You must make sure that you show the correct number of zeros.
- If you round the number based on the first digit, you'll round 9 to 10. You write 1000, not 100.

970 rounds to 1000

- Remember, if the 9 rounds up, write 10, but make sure you have the correct number of zeros.

Part 5

For each item, write the rounded value that is based on the first digit.

a. 2978 b. 93,400 c. 9740 d. 968 e. 8903 f. 47

Part 6

Make a ratio table. Answer the questions in each item. Write your answer as a number and a unit name.

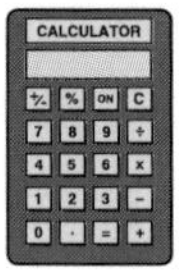

a. Bernard has 2 U.S. stamps for every 7 foreign stamps.
   1. If Bernard has 105 foreign stamps, how many U.S. stamps does he have?
   2. How many stamps does he have in all?

b. In the school library, there are hardback books and paperback books. The ratio of hardback books to total books is 5 to 9. There is a total of 720 books in the library.
   1. How many hardback books are there?
   2. How many paperback books are there?

c. The animals Dr. Jones counted were squirrels and foxes. For each fox he counted, there were 6 squirrels. Dr. Jones counted 95 foxes.
   1. How many squirrels were counted?
   2. How many animals did Dr. Jones count in all?

Part 7 **For each picture, make a fraction number family. Show the groups being removed as the first small number in your family.**

a. 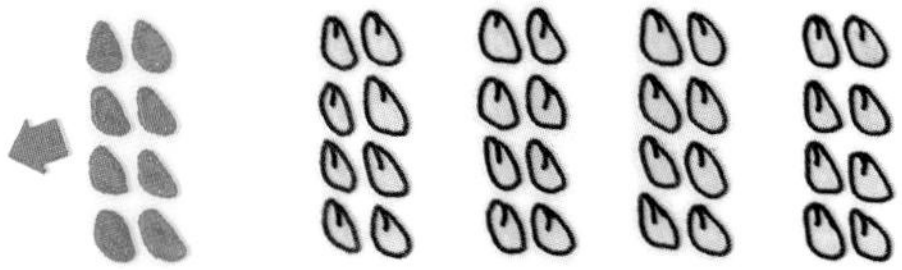

b. 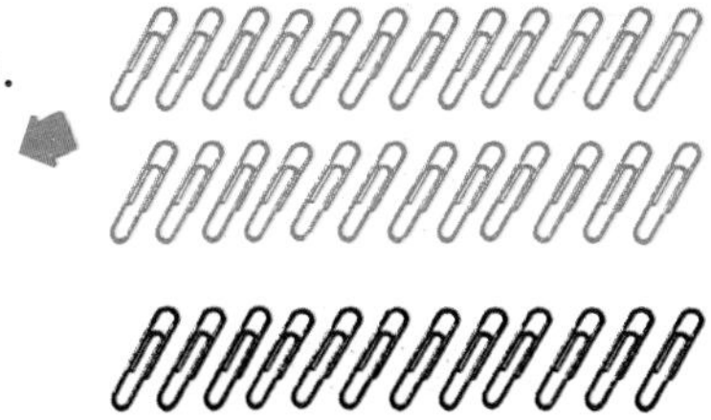

c. 

d. 

Independent Work

Part 8 **For each item, make a fraction number family. Box the fraction that answers the question.**

a. $\frac{2}{11}$ of the chickens did not lay eggs. What fraction of the chickens laid eggs?

b. In a school, some students wore shoes that had rubber soles. $\frac{7}{10}$ of the students did not wear shoes with rubber soles. What fraction of the students wore shoes with rubber soles?

c. $\frac{3}{5}$ of the children in a choir were boys. What fraction of the children were girls?

Part 9 **Write the degrees for each lettered angle.**

Angle t is 74°.

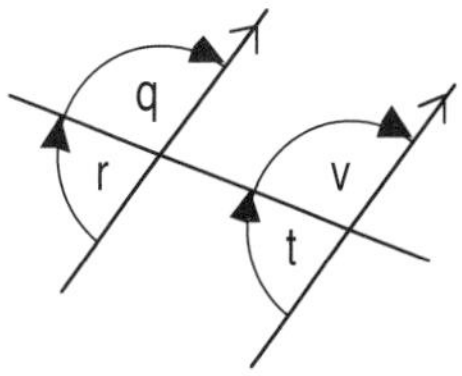

Part 10 **Copy and complete each equation.**

a. $\frac{184}{9} = ■$

b. $\frac{8}{3} - \frac{2}{3} = ■$

c. $874 \times ■ = 0$

d. $\frac{13}{4}(■) = \frac{13}{12}$

e. $5 = \frac{■}{10}$

f. $\frac{9}{10} \times \frac{6}{10} = ■$

g. $\frac{3}{7} = \frac{■}{42}$

h. $■ - 74 = 1$

i. $\frac{4}{8} - \frac{0}{8} + \frac{2}{8} = ■$

j. $■ + \frac{4}{5} = \frac{5}{5}$

Part 11 **Copy each problem and multiply. If the answer is too big or too small, work the problem again with the correct answer.**

a. $66\overline{)480}$ with 6 written above

b. $73\overline{)634}$ with 9 written above

c. $59\overline{)444}$ with 7 written above

Part 12 Answer each question.

a. A pie was divided into 6 pieces that were the same weight. Each piece weighed 25 grams. How much did the entire pie weigh?

b. A pie was divided into 7 pieces that were the same weight. The entire pie weighed 210 grams. How much did each piece weigh?

c. A pie was divided into pieces that were the same weight. The entire pie weighed 300 grams. Each piece weighed 30 grams. How many pieces were there?

d. A pie was divided into pieces that were the same weight. Each piece weighed 40 grams. There were 6 pieces. How much did the entire pie weigh?

Part 13 Copy the table and complete all the rows.

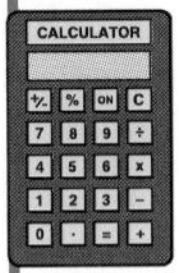

	Multiplication	Division	Fraction equation
a.	14 (■) = 70		
b.		$13\overline{)78}$	
c.			$\frac{420}{21}$ = ■

Part 14 Copy and complete each table.

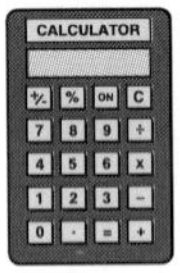

a. The names for the table are cars, vehicles and vehicles that are not cars. The number for cars in the first column is 3.

	5	■
	■	51
	8	■

b. The names for the table are dogs, long-tailed dogs and short-tailed dogs. The number for long-tailed dogs in the first column is 2.

	13	■
	■	■
	15	45

Part J

a. A□ = b×h
A□ = 15×8
A□ = 120 sq m

8
8
15
+15
P = 46 m

b. A△ = $\frac{b \times h}{2}$
A△ = $\frac{14 \times 12}{2}$ = $\frac{168}{2}$
A△ = 84 sq ft

15
13
+14
P = 42 ft

c. A□ = b×h
A□ = 7×8
A□ = 56 sq m

10
10
7
+7
P = 34 m

Part K

a. $19\overline{)94}$
−76
18
$4\frac{18}{19}$

b. $82\overline{)501}$
−492
9
$6\frac{9}{82}$

c. $68\overline{)499}$
−476
23
$7\frac{23}{68}$

Lesson 40

Part 1 **Use an estimate to work each division problem.**

a. $23\overline{)187}$ b. $67\overline{)340}$ c. $88\overline{)538}$

Test 4

Part 1 **Make a table and answer the questions.**

There were houses in Villa Park and Oak Grove. Some had basements. Some didn't. There were 27 houses with basements in Villa Park and 54 houses with basements in Oak Grove. The total number of houses that did not have basements was 148. The total number of houses in Villa Park was 115.

a. How many houses were in Oak Grove?

b. What was the total number of houses in both communities?

c. Were there more houses without basements in Villa Park or Oak Grove?

d. How many total houses had basements?

Part 2 Find the area and perimeter of each figure.

a.

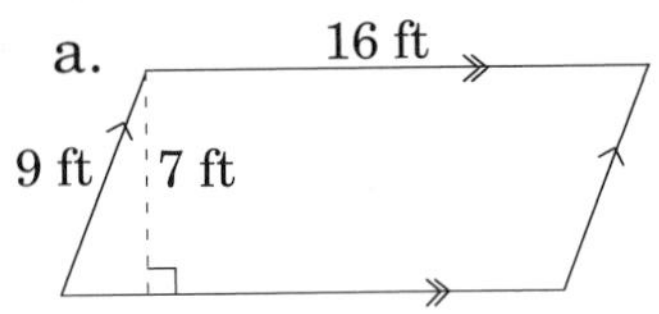

b.

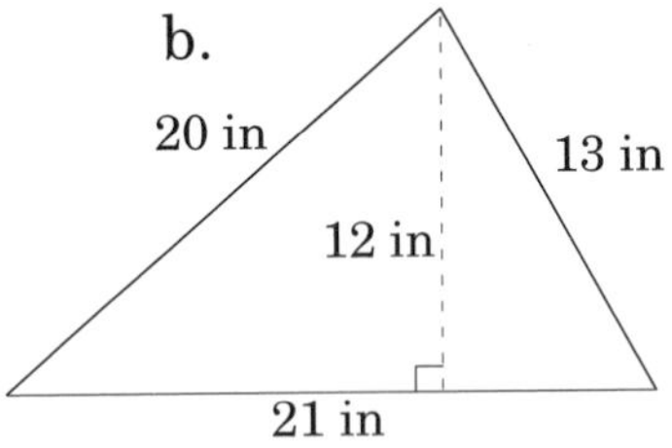

c.

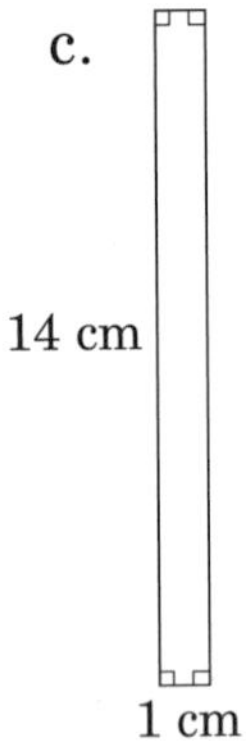

Part 3 For each item, make a fraction number family with three names and three fractions. Box the fraction that answers the question.

a. $\frac{3}{8}$ of the children had sleeping bags. What fraction of the children did not have sleeping bags?

b. $\frac{4}{7}$ of the doors in a building were closed. What fraction of the doors were open?

Part 4 For each item, trace the figure and write the degrees for all lettered angles.

a. Angle k is 115°.

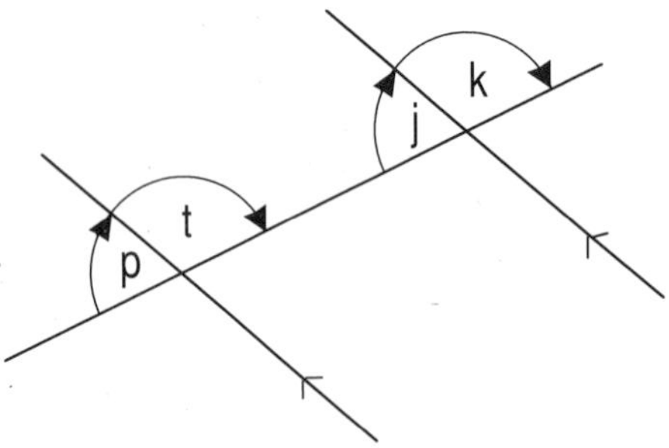

b. Angle g is 75°.

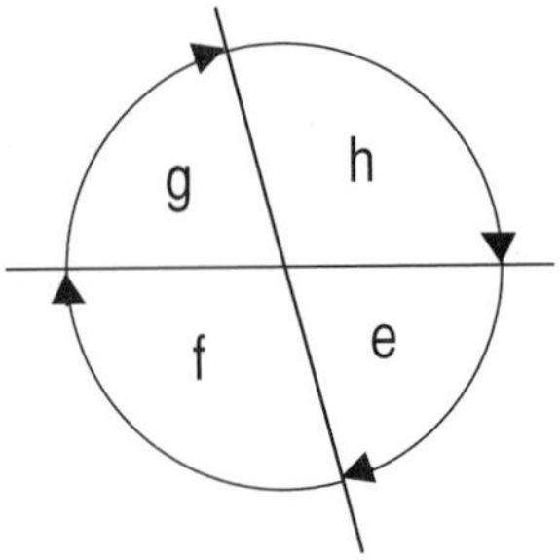

Part 5

For each problem, make a ratio table. Answer the questions.

a. At a concert, there are 7 women to every 9 adults. There are 800 men.
   1. How many women are there?
   2. How many adults are there?

b. On a long trip, the ratio of cloudy days to sunny days was 1 to 5. The trip lasted 48 days.
   1. How many days were sunny?
   2. How many days were cloudy?

Part 6

If the number shown in the answer is correct, work the problem. If the number shown in the answer is not correct, use the correct number and work the problem.

a. $17\overline{)135}$ with 8 shown as the answer

b. $42\overline{)218}$ with 4 shown as the answer

Part 7

For each item, write an equation. Show the fraction and the whole number or mixed number it equals.

a. $\frac{108}{6}$ b. $\frac{108}{7}$ c. $\frac{108}{4}$ d. $\frac{108}{5}$

Part 8

For each item, make a number family and answer the question.

a. The tree was 41 meters taller than the pole. The tree was 47 meters tall. What was the height of the pole?

b. In a pond, there were 57 fewer ducks than other birds. There were 98 ducks. How many other birds were there?

c. 134 of the workers in a factory wore gloves. 307 workers did not wear gloves. How many workers were there in the factory?

Lesson 41

Part 1 **For each picture, make a fraction number family. Show the parts being removed as the first small number in your family.**

a. 

b. 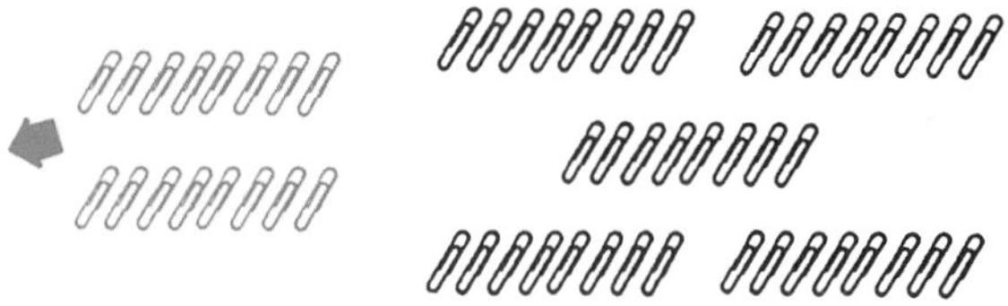

c. 

d. 

Part 2 **Find the area and the perimeter of each figure.**

a. 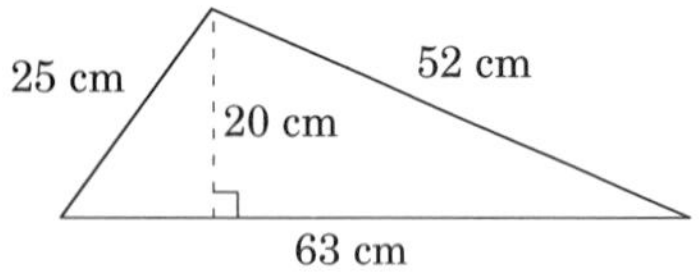

b. 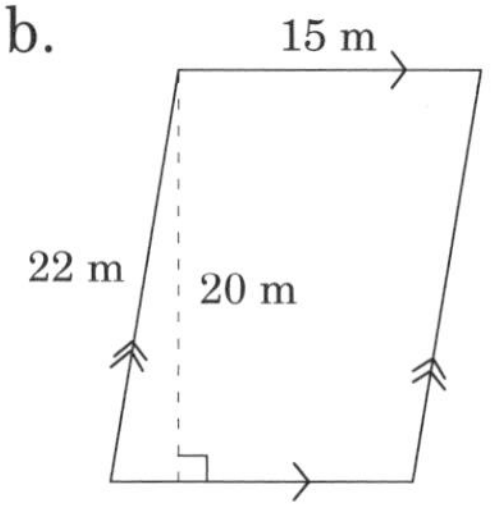

c. 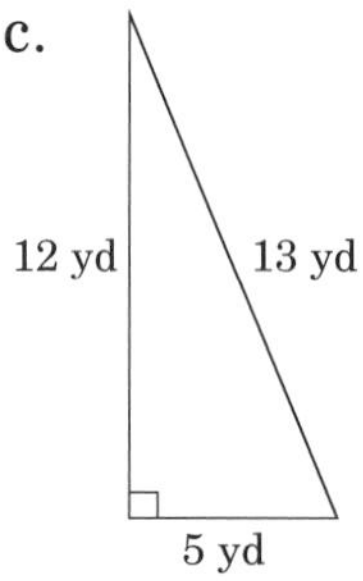

Part 3 **For each item, make a fraction number family. Then make a ratio table and answer the questions.**

Sample problem

$\frac{2}{5}$ of the workers in a company took a bus to work. The rest didn't. There was a total of 745 workers in the company.

1. How many workers took a bus to work?
2. How many workers didn't take a bus to work?

a. There were new and used cars on a lot. $\frac{3}{7}$ of the cars were used. There were 64 new cars on the lot.

1. How many used cars were on the lot?
2. How many total cars were on the lot?

b. A laboratory made a mixture of sulphur and water. $\frac{4}{5}$ of the mixture was sulphur. The rest was water. The total mixture weighed 75 pounds.

1. How many pounds of sulphur were in the mixture?
2. How many pounds of water were in the mixture?

c. $\frac{5}{8}$ of the corn harvest is ground into cornmeal. The rest is canned. This year 93 pounds of corn were canned.

1. How much corn was made into cornmeal?
2. What was the total corn harvest?

Part 4

For each item, write an estimation problem and the answer.

a.
$$\begin{array}{r} 496 \\ 9983 \\ +\ 222 \\ \hline \end{array}$$

b.
$$\begin{array}{r} 48 \\ 96 \\ +\ 2038 \\ \hline \end{array}$$

c.
$$\begin{array}{r} 2430 \\ 997 \\ +\ 5670 \\ \hline \end{array}$$

Part 5

- You know how to work multiplication problems that have a missing middle value. You work a division problem.

  $4(\blacksquare) = 20$ $\qquad$ $4\overline{)20}$

- You can also write the missing value as a fraction. You just write the fraction for the division problem.
- Here's the fraction for 20 divided by 4: $\frac{20}{4}$
- That's the missing value. $4\left(\frac{20}{4}\right) = 20$
- Remember, say the division problem, then write the fraction for that problem.

Part 6 Copy each problem and write the missing value as a fraction.

a. $3(\blacksquare) = 11$ b. $5(\blacksquare) = 6$ c. $9(\blacksquare) = 20$

d. $6(\blacksquare) = 50$ e. $2(\blacksquare) = 19$ f. $7(\blacksquare) = 40$

Part 7

- You can write equations that show the degrees for angles.
- You'll use this symbol to identify a lettered angle: $\angle$
- Here's an equation that says, **Angle J equals 61 degrees:** $\angle j = 61°$

a. $\angle f = 108°$

b. $\angle n = 37°$

Part 8 Write an equation to show the degrees for the lettered angle.

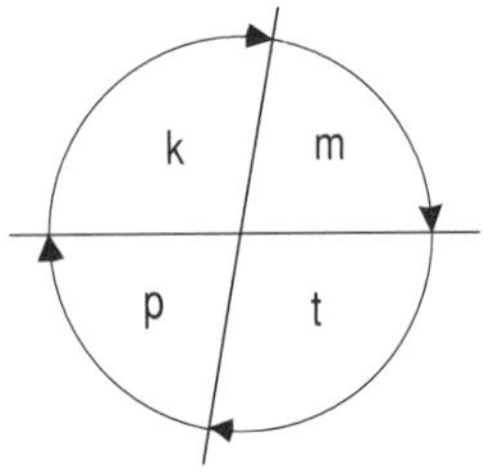

$\angle m = 80°$

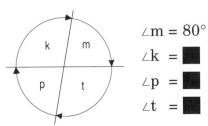

$\angle k = \blacksquare$

$\angle p = \blacksquare$

$\angle t = \blacksquare$

Independent Work

Work the division problems for the fractions in part 6.

Part 9 Make a table and fill in all the numbers. Then answer the questions.

Two lakes were stocked with trout and perch. The lakes were Crystal Lake and Diamond Lake. 317 trout were stocked in Diamond Lake. The total number of fish stocked in Diamond Lake was 484. The total number of perch for both lakes was 460. The total number of fish in both lakes was 841.

a. How many perch were put in Diamond Lake?

b. How many trout were put in Crystal Lake?

c. Were more fish put in Diamond Lake or Crystal Lake?

Part 10 **Copy and complete the table.**

	Division	Fraction equation
a.	$5\overline{)48}$	$\frac{48}{5}$ = ■
b.	$7\overline{)347}$	■ = ■
c.	$5\overline{)555}$	
d.	$3\overline{)306}$	

Part 11 **For each problem, make a number family. Answer each question with a number and unit name.**

a. The mountains are 19 miles closer than the beach. The mountains are 74 miles away. How far away is the beach?

b. Fred weighs 124 pounds. His horse weighs 1340 pounds. How much lighter is Fred than his horse?

c. In Marion County there are 779 pet birds. There are 449 pet dogs. How many fewer pet dogs than pet birds are there in Marion County?

d. The deli sold 128 fewer cookies than the bakery. The bakery sold 210 cookies. How many cookies did the deli sell?

Part 12 **Copy and work each problem.**

a. $34\overline{)163}$ b. $46\overline{)238}$ c. $53\overline{)272}$ d. $28\overline{)114}$

Part 13 **For each item, make a table and complete it.**

a. Multiply each number by 2. Then add 5.
0, 20, 1, 3

b. Add 2 to each number. Then multiply the sum by 5.
0, 20, 1, 3

Part 14 **For each item, write a ratio equation and answer the question.**

a. In a park, there were 9 people for every 2 benches. There were 54 people. How many benches were there?

b. Every 4 bricks weighed 9 pounds. There were 36 pounds of bricks. How many bricks were there?

Part 15 **For each item, make a fraction number family. Box the answer.**

a. $\frac{2}{41}$ of the shirts had flaws. What fraction of the shirts did not have flaws?

b. $\frac{7}{15}$ of the vehicles did not have radial tires. What fraction of the vehicles did have radial tires?

Part J

a. $A\triangle = \frac{b \times h}{2}$ $A\triangle = \frac{63 \times 20}{2} = \frac{1260}{2}$ $A\triangle = 630$ sq cm

25 52 +63 P=140 cm

b. $A\square = b \times h$ $A\square = 15 \times 20$ $A\square = 300$ sq m

22 22 15 +15 P=74 m

c. $A\triangle = \frac{b \times h}{2}$ $A\triangle = \frac{5 \times 12}{2} = \frac{60}{2}$ $A\triangle = 30$ sq yd

12 13 +5 P=30 yd

Lesson 42

Part 1

- You've worked with fractions that have a denominator of 1. Any number over 1 equals the number.

  $\frac{5}{1} = 5$ $\quad \frac{243}{1} = 243$

- You can show that these fractions equal the whole number by using your calculator.

- To do that, you work the division problem for the fraction.

  $\frac{■}{■} = ■ \boxed{\div} ■ \boxed{=}$

a. $\frac{34}{1}$ b. $\frac{3}{1}$

Part 2 Copy and work each problem.

a. $\frac{5}{2} \times \frac{1}{2} \times \frac{3}{4} =$ ■

b. $2 \times 6 \times 7 =$ ■

c. $2 \times \frac{3}{2} \times \frac{5}{8} =$ ■

Part 3 Copy each problem and write the missing value as a fraction. Below, rewrite the equation using the whole number.

a. $8(■) = 648$

b. $7(■) = 434$

c. $3(■) = 1503$

Part 4

- All the values that are written before the decimal point are 1 or more.
- All values that come after the decimal point are less than 1.

- You can write a decimal point after any whole number. That's what your calculator does.

  203. 1. 45.

- The columns show how whole numbers and decimals work.
- Whole numbers between 1000 and 9999 have four digits. The first digit is the thousands digit.
- Whole numbers between 100 and 999 have three digits. The first digit is the hundreds digit.
- Whole numbers between 10 and 99 have two digits. The first digit is the tens digit.
- Whole numbers between 1 and 9 have one digit. That digit tells the number of ones.
- If the decimal part ends one place after the decimal point, that part tells about tenths.
- If the decimal part ends two places after the decimal point, that part tells about hundredths.
- If the decimal part ends three places after the decimal point, that part tells about thousandths.

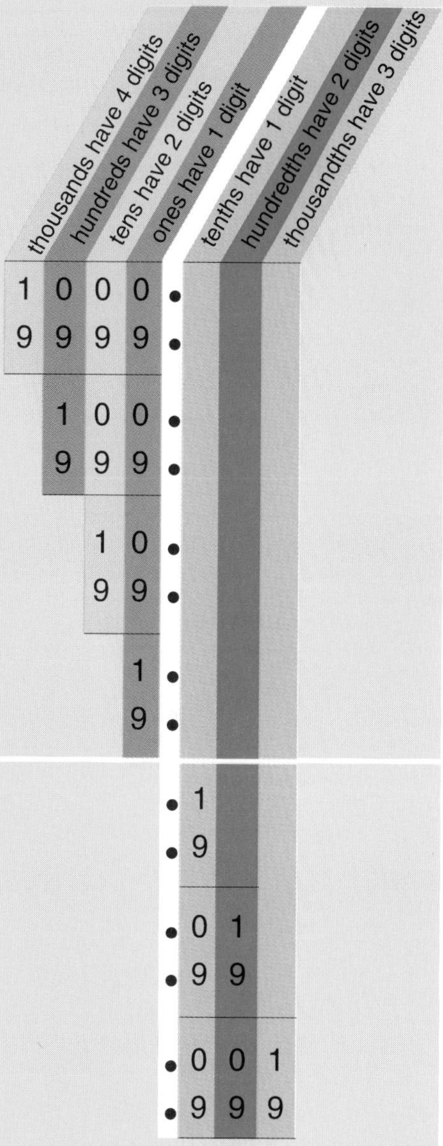

Part 5

Here are rules for reading decimal numbers:

- You read the whole number part. You say **and** for the decimal point. Then you read the decimal part.
- If the number ends **one** place after the decimal point, you say **tenths.**
- If the number ends **two** places after the decimal point, you say **hundredths.**
- If the number ends **three** places after the decimal point, you say **thousandths.**

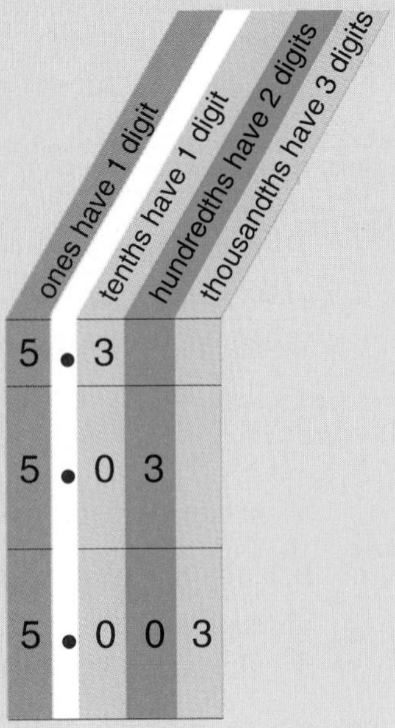

Part 6

Read each decimal number.

a. 6.35
b. 5.13
c. 6.4
d. 5.3
e. 7.41
f. 25.6
g. 3.189
h. 18.502

Part 7

For each problem, make a ratio table and answer the questions.

a. $\frac{4}{9}$ of the workers in a factory wore glasses. 76 workers wore glasses.
   1. How many of the workers did not wear glasses?
   2. How many workers were there in all?

b. The ratio of sand to cement in a mixture is 7 to 2. The mixture weighs 342 pounds.
   1. How many pounds of sand are in the mixture?
   2. How many pounds of cement are in the mixture?

c. $\frac{3}{8}$ of the students in the school caught chicken pox. 125 students did not catch chicken pox.
   1. How many students caught chicken pox?
   2. How many students were there in all?

Part 8 Answer each question.

a. 1. What is the fraction for the whole pie?
   2. What is the fraction for 2 pieces of the pie?

b. 1. What is the fraction for 8 pieces?
   2. What is the fraction for 1 piece?

c. 1. What is the fraction for 2 groups?
   2. What is the fraction for 3 groups?

d. 1. What is the fraction for 1 group?
   2. What is the fraction for 6 groups?

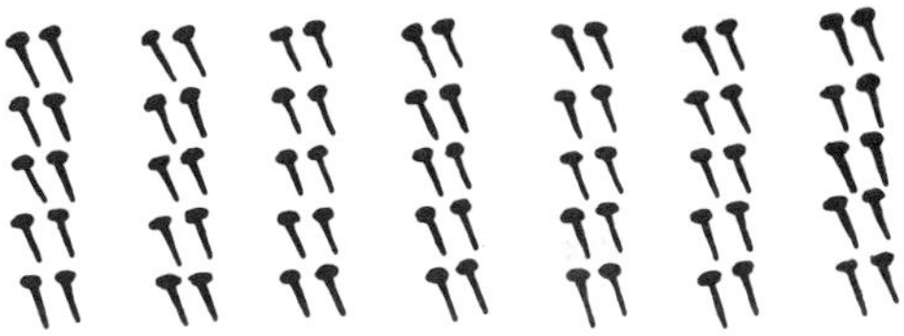

Independent Work

Part 9 For each item, write a ratio equation. Answer the question with a number and a unit name.

a. A factory shipped 4 boxes every 3 minutes. How many minutes would it take the factory to ship 164 boxes?

b. Every 3 cows produced 11 gallons of milk. How many cows were needed to produce 550 gallons of milk?

c. 7 pounds of dirt were removed from the pile every 9 seconds. How long would it take for 504 pounds of dirt to be removed from the pile?

Part 10 Copy and work each problem.

a. $29\overline{)286}$ b. $23\overline{)169}$ c. $54\overline{)277}$

Part 11

Find the area and perimeter of each figure.

a.

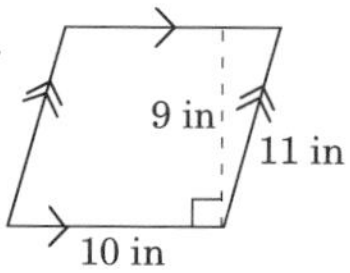

b.

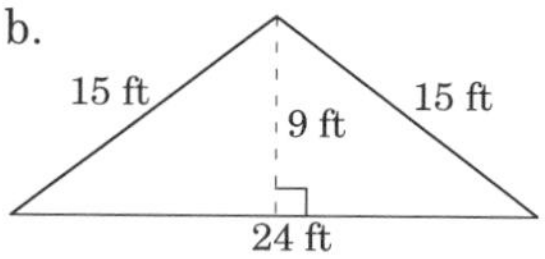

c.

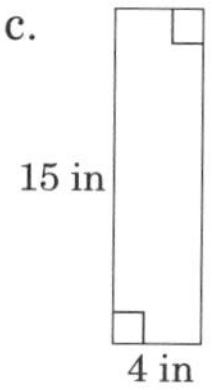

Can you find the area of that large triangular lot ?
Sure, it's easy to find. It's right over there.

Part 12

Copy and complete the equation for each lettered angle.

v = 130°

z = ■

x = ■

y = ■

r = ■

t = ■

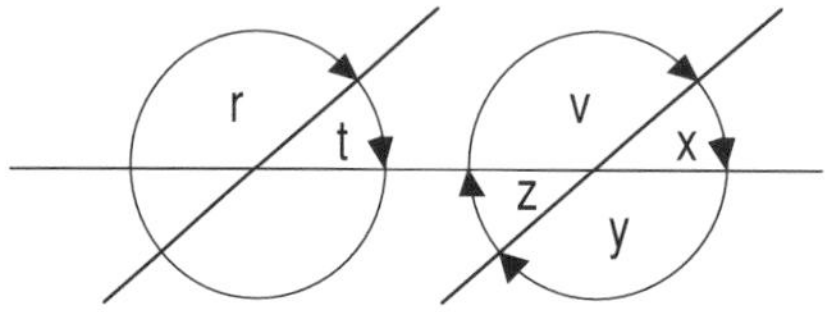

Part 13

Rewrite each equation with the letter on the left. Below, write what each letter equals.

a. 436 – 111 = M

b. 56 + 491 = P

c. $\frac{84}{2}$ = R

Part 14

Copy each fraction and write the mixed number it equals.

a. $\frac{74}{3}$ b. $\frac{197}{9}$ c. $\frac{197}{8}$

Part 15

Copy each mixed number and write the fraction it equals.

a. $7\frac{3}{9}$ b. $31\frac{4}{10}$ c. $3\frac{42}{65}$

Lesson 43

Part 1 **Copy each family. Write the missing fraction.**

a. $\frac{3}{20} \longrightarrow \blacksquare \longrightarrow \frac{25}{20}$

b. $\frac{4}{9} \quad \frac{12}{9} \longrightarrow \blacksquare$

c. $\frac{3}{9} \longrightarrow \blacksquare \longrightarrow \frac{12}{9}$

d. $\frac{1}{10} \longrightarrow \blacksquare \longrightarrow \frac{50}{10}$

Part 2 **Copy and work each problem.**

a. $\frac{3}{4} \times 5 \times \frac{2}{5} = \blacksquare$

b. $\frac{1}{3} \times \frac{5}{6} \times \frac{7}{2} = \blacksquare$

c. $2 \times 5 \times 7 = \blacksquare$

Part 3 **For each problem, make a ratio table and answer the questions.**

a. In a cake recipe, there are 2 cups of water for every 5 cups of milk. The combined amount of milk and water used was 84 cups.
   1. How many cups of water were used?
   2. How many cups of milk were used?

b. $\frac{5}{7}$ of the people on vacation wore ski suits. 95 people wore ski suits.
   1. How many people were on vacation?
   2. How many people did not wear ski suits?

c. On $\frac{1}{5}$ of the school days, the bus arrived late. There were 210 days in all.
   1. On how many days did the bus arrive late?
   2. On how many days did the bus arrive on time?

Part 4 Answer each question.

a. 1. What is the fraction for 2 groups?
   2. What is the fraction for 5 groups?

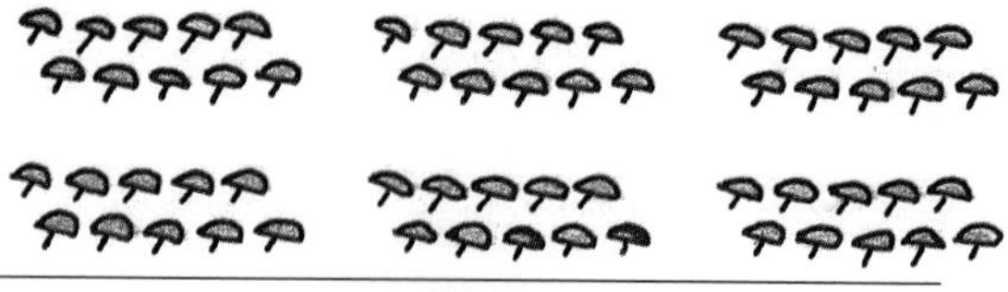

---

b. 1. What is the fraction for 6 pieces?
   2. What is the fraction for 1 piece?

---

c. 1. What is the fraction for 4 groups?
   2. What is the fraction for 1 group?

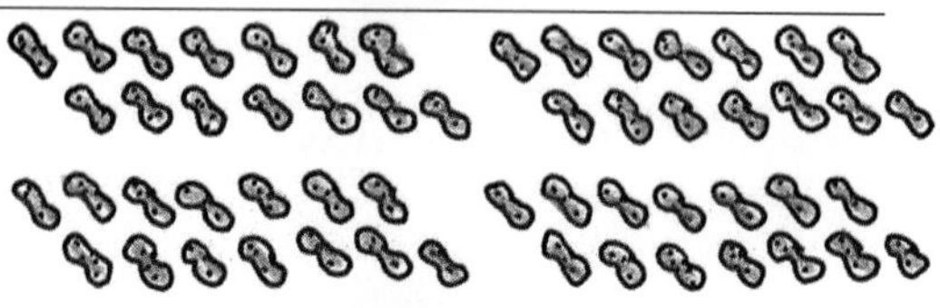

---

d. 1. What is the fraction for 6 pieces?
   2. What is the fraction for 3 pieces?

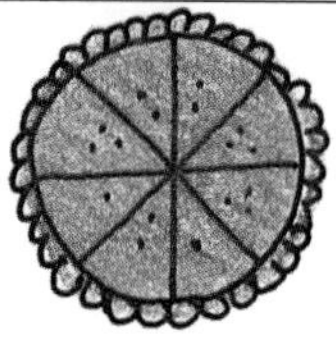

Part 5 Read each fraction as a division problem.

a. $\frac{3}{4}$ b. $\frac{1}{7}$ c. $\frac{24}{39}$

Part 6 Copy each problem and write the missing value as a fraction.

a. $21(\blacksquare) = 5$
b. $18(\blacksquare) = 9$
c. $7(\blacksquare) = 11$
d. $19(\blacksquare) = 4$
e. $50(\blacksquare) = 17$
f. $39(\blacksquare) = 75$

Independent Work

Part 7 **For each problem, make a number family. Answer the questions.**

a. There were 285 more people registered to vote in 1989 than in 1988. In 1989, there were 1572 people registered. How many people were registered to vote in 1988?

b. 111 of the children had been vaccinated. The rest hadn't. There were 131 children. How many had not been vaccinated?

c. The apartment complex is 128 feet higher than the office building. The office building is 391 feet high. How high is the apartment complex?

d. 48 of the peanuts in a bag had been eaten. 358 had not been eaten. How many peanuts had been in the bag to begin with?

Part 8 **For each problem, make a fraction number family. Box the fraction that answers each question.**

a. $\frac{2}{10}$ of the questions were difficult. What fraction were not difficult?

b. On Sunday, it was not raining $\frac{3}{12}$ of the time. What fraction of the time was it raining on Sunday?

Part 9 **Copy and work each problem.**

a. $59\overline{)206}$

b. $38\overline{)308}$

c. $77\overline{)555}$

Part 10 **Complete the equation for each lettered angle.**

a. $\angle h = 52°$
$\angle f = \blacksquare$
$\angle d = \blacksquare$
$\angle g = \blacksquare$

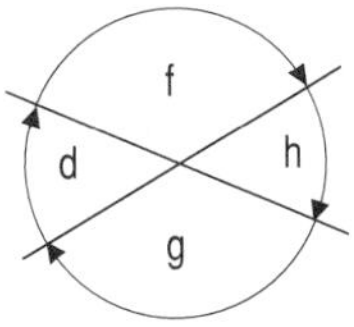

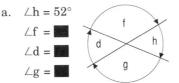

b. $\angle p = 115°$
$\angle f = \blacksquare$
$\angle m = \blacksquare$
$\angle s = \blacksquare$

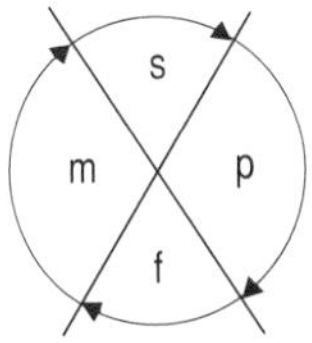

Part 11 **Copy and work each problem.**

a. $17 = \frac{\blacksquare}{20}$

b. $3 = \frac{\blacksquare}{50}$

c. $\frac{4}{5} = \frac{80}{\blacksquare}$

d. $\frac{1}{7} = \frac{100}{\blacksquare}$

e. $\frac{4}{8} \times \frac{1}{3} \times \frac{3}{2} = \blacksquare$

f. $\frac{5}{8} - \frac{1}{8} - \frac{3}{8} = \blacksquare$

g. $\frac{2}{5}(\blacksquare) = \frac{24}{25}$

h. $\frac{1}{8} \times \blacksquare = \frac{20}{40}$

Part 12 **Find the area and the perimeter of each figure.**

a.

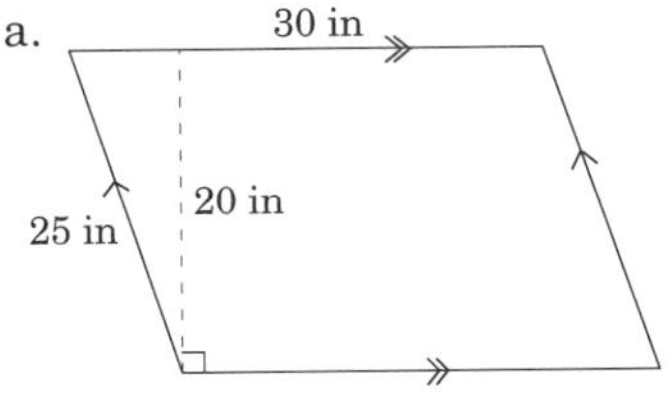

b.

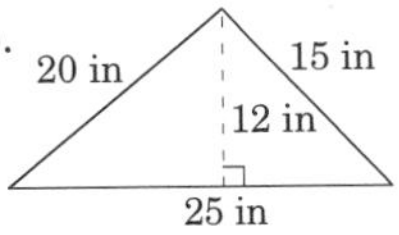

Lesson 44

Part 1 **For each item, make a fraction number family. Box the answer to the question.**

Sample
A pie was divided into 6 equal parts. A person removed 4 parts. What is the fraction for the pie that remains?

a. Seeds were divided into 12 equal-sized groups. 7 of those groups were removed. What's the fraction for the groups that were not removed?

b. A lot was divided into 5 equal-sized parcels. 4 of the parcels were sold. What fraction of the parcels were not sold?

c. A large roast was cut into 13 pieces that were the same weight. 3 of the pieces were eaten. What's the fraction for the pieces that were not eaten?

d. A person divided her money into 4 equal parts. 2 parts went to taxes. What fraction of the person's money did not go to taxes?

Part 2

- You're going to work with prime numbers.
- Here's a rule about prime numbers:

  If you divide a **prime** number by anything other than 1 or the number itself, you won't get a whole-number answer.

- **5** is a prime number because it can't be divided evenly by anything but 1 or 5.
- **7** is a prime number because it can't be divided evenly by anything but 1 or 7.
- **16** is **not** a prime number because it **can** be divided by something other than 1 and 16. It can be divided by 2, or by 4, or by 8.

Prime numbers

2, 3, 5, 7, 11, 13, 17, 19, 23, 29, 31, 37

- **Prime numbers that are multiplied together are prime factors.**
- You can show numbers that are not primes as prime factors.
- 45 is not a prime number. You can show it as prime factors. **45**
- You start with any two factors of 45: **45 = 9 x 5**
- 9 is not a prime. So you rewrite 9 as prime factors. **45 = 3 x 3 x 5**

Sample items

a. **8**
8 = 2 x 4
8 = ■ x ■ x ■

b. **48**
48 = 6 x 8
48 = ■ x ■ x ■ x ■ x ■

Part 3

For each item that doesn't show the prime factors, rewrite the bottom equation to show the prime factors multiplied together.

a. 12 = 4 x 3
12 = 2 x 2 x 3

b. 50 = 10 x 5

c. 40 = 2 x 4 x 5

d. 54 = 6 x 9
54 = 2 x 3 x 9
54 = 2 x 3 x 3 x 3

e. 70 = 2 x 35

Part 4 **For each item, write the name or letter of the larger thing.**

> ***Rules***
>
> - If the fraction is **more than 1,** the first thing named is **larger.**
> - If the fraction is **less than 1,** the first thing named is **smaller.**

Sample 1 Jan is $\frac{5}{8}$ as tall as Fran.

Sample 2 Jan is $\frac{8}{7}$ the height of Ted.

Sample 3 Pile A is $\frac{7}{5}$ the weight of pile B.

a. Hilda's weight is $\frac{9}{8}$ of Edna's weight.

b. Pile C is $\frac{4}{5}$ the weight of pile M.

c. Pile J is $\frac{7}{3}$ the weight of pile B.

d. The turtle's age was $\frac{8}{7}$ the age of the tree.

e. The area of the farm was $\frac{13}{14}$ the area of the forest.

Part 5 **For each item, make the number family. Rewrite the fraction that equals 1 with the correct denominator. Then write the missing fraction.**

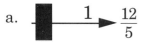

a. $\blacksquare \xrightarrow{\quad 1 \quad} \frac{12}{5}$

b. $\blacksquare \xrightarrow{\quad 1 \quad} \frac{17}{15}$

c. $\blacksquare \xrightarrow{\quad \frac{6}{20} \quad} 1$

d. $\blacksquare \xrightarrow{\quad 1 \quad} \frac{30}{4}$

e. $\blacksquare \xrightarrow{\quad \frac{9}{10} \quad} 1$

Independent Work

Part 6 **Write the estimation problem and the estimation answer. Rework any problem with the wrong answer.**

a. $\begin{array}{r} 294 \\ 993 \\ +\ 115 \\ \hline 1302 \end{array}$

b. $\begin{array}{r} 527 \\ -\ 390 \\ \hline 237 \end{array}$

c. $\begin{array}{r} 1964 \\ -\ 971 \\ \hline 893 \end{array}$

d. $\begin{array}{r} 483 \\ 780 \\ +\ 96 \\ \hline 1129 \end{array}$

Part 7 Make a table. Answer each question.

Hard wheat and soft wheat were grown on different farms—Brown Farm and Roger Farm. Brown Farm produced 680 tons of soft wheat. Roger Farm produced 295 tons of hard wheat. The total amount of wheat produced on Roger Farm was 450 tons. The total amount of wheat produced on both farms was 1265 tons.

a. What is the total amount of hard wheat grown on both farms?

b. Was more hard wheat grown on Brown Farm or on Roger Farm?

c. Was less hard wheat or soft wheat grown on Roger Farm?

Part 8 Find the area and perimeter of each figure.

a. 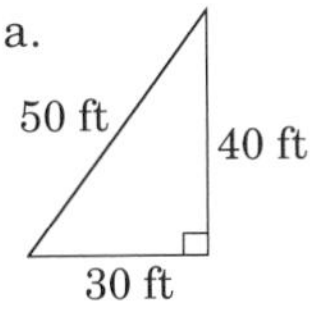

b. 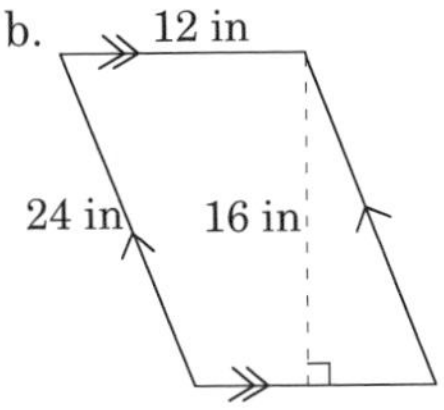

c. 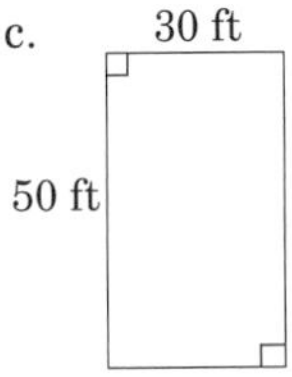

Part 9 Copy and work each problem.

a. $\frac{2}{9} + \frac{9}{9} =$ ■

b. $\frac{1}{3} \times \frac{9}{7} =$ ■

c. $\frac{11}{4} - \frac{10}{4} =$ ■

d. $\frac{3}{5} \times \frac{9}{5} =$ ■

e. $\frac{2}{3} \times 5 \times \frac{1}{2} =$ ■

f. $2 \times \frac{1}{2} \times 3 =$ ■

g. $\frac{5}{3} \times \frac{2}{5} \times 4 =$ ■

Part 10 Figure out the missing angles marked with a letter.

a. 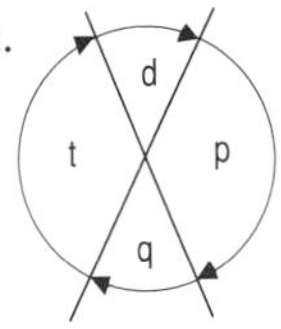

$\angle d = 49°$

$\angle p =$ ■

$\angle q =$ ■

$\angle t =$ ■

b. 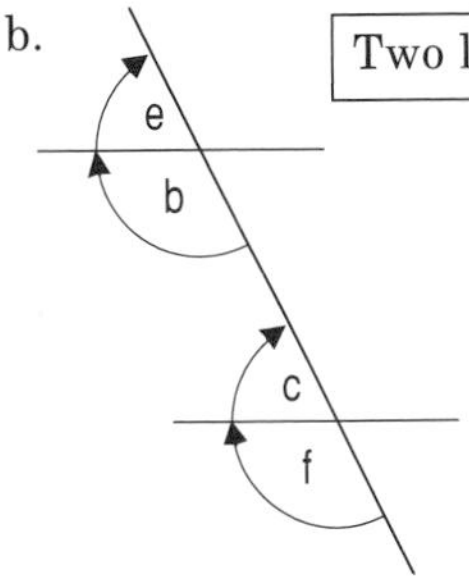

Two lines are parallel.

$\angle f = 117°$

$\angle e =$ ■

$\angle b =$ ■

$\angle c =$ ■

Part 11 For each item, make a table and complete it.

a. Add 4 to each number.
Then multiply the sum by 4.
1, 2, 4

b. Subtract 5 from each number.
Then multiply by 2.
6, 25, 12, 5

Part 12 Copy and complete each ratio table.

a.

■	■
8	64
13	■

b.

■	105
4	■
11	■

Lesson 45

Part 1

- You can write fractions for decimal numbers. You just read the decimal number and write the fraction for what you read.
- Here's: **.35** You write the fraction: $\frac{35}{100}$
- Here's: **.002** You write the fraction: $\frac{2}{1000}$
- Here's: **.2** You write the fraction: $\frac{2}{10}$
- Remember, when you read the decimal value, you say the fraction you'll write.

Part 2

Copy the table. Fill in the missing values.

	Decimal	Fraction
a.	.5	
b.	.8	
c.	.08	
d.	.003	
e.	.03	

Part 3

For each item, make a fraction number family and box the fraction that answers the question.

a. A pizza was divided into 9 equal pieces. 7 pieces had onions on them. What fraction of the pizza did not have onions?

b. The workers in a factory formed 11 baseball teams. 3 of those teams won more than 10 games. What fraction of the teams did not win more than 10 games?

c. A truckload of sand was divided into 17 piles of equal weight. 5 of the piles were used to make concrete. 12 piles were not used. What fraction of the sand was not used?

d. At Hope Hospital, 145 babies were born in 1988. 77 of those babies were boys. What fraction of the babies were girls?

Part 4 For each item, write an equation that shows only prime factors.

a. 56 = 8 x 7   b. 30 = 5 x 6   c. 28 = 7 x 4   d. 54 = 6 x 9

Part 5 For each item, write the name or letter of the thing that is more.

a. The cost of the couch was $\frac{3}{7}$ as much as the cost of the bed.

b. The dog was $\frac{6}{5}$ as old as the cat.

c. Mary was $\frac{7}{9}$ as tall as Fran.

d. Jim earned $\frac{9}{7}$ the amount that Dan earned.

e. Orville had $\frac{2}{3}$ the number of cards that Jay had.

Part 6 Make a table. Then put in the numbers the problem gives, figure out the missing numbers and write answers to the questions.

There were two crews of workers—the green crew and the red crew. In each crew there were experienced workers and inexperienced workers. The total number of inexperienced workers for both crews was 101. There were 48 experienced workers on the green crew. There were 70 inexperienced workers on the red crew. The total number of workers on the red crew was 121.

Questions

a. Which crew had more inexperienced workers?

b. How many total workers were on the green crew?

c. What was the total number of experienced workers?

d. What was the total number of workers for both the crews?

e. How many inexperienced workers were on the green crew?

Independent Work

Part 7 **Make a fraction number family. Complete a ratio table and answer the questions.**

$\frac{5}{7}$ of the shoes were sneakers. 84 shoes were not sneakers.

a. How many shoes were there?

b. How many sneakers were there?

Part 8 **Copy and work each problem.**

a. $3\overline{)4569}$

b. $2\overline{)1381}$

c. $4\overline{)2104}$

d. $78\overline{)253}$

e. $63\overline{)520}$

Part 9 **Figure out all the missing angles.**

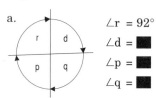

a.

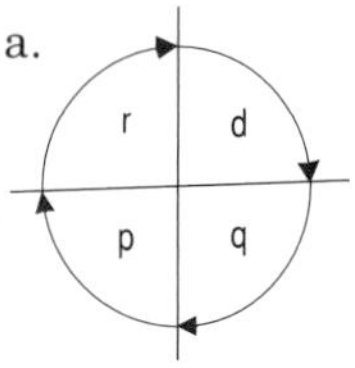

∠r = 92°
∠d = ■
∠p = ■
∠q = ■

b.

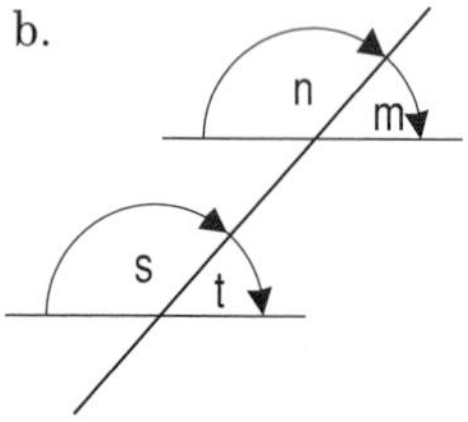

Two lines are parallel.

∠m = 48°
∠s = ■
∠n = ■
∠t = ■

Part 10 **Answer each question.**

a. In a pond, there were 7 tadpoles for every 6 minnows. The pond had a total of 618 minnows. How many tadpoles were in the pond?

b. A truck delivered 16 tons of concrete every 3 hours. How many hours would it take for the truck to deliver 96 tons of concrete?

c. In a school, there were 3 girls for every 4 boys. The school had 200 boys. How many girls were in the school?

Part 11 **Copy the table and complete all the rows.**

	Multiplication	Division	Fraction equation
a.	9 (■) = 324		
b.	2 (■) = 56		
c.	5 (■) = 55		

Part 12 **For each item, make a table and complete all the rows.**

a. Add 10 to each number.
Subtract the sum from 100.
23, 5, 10

b. Multiply each number by 11.
Add 3 to each product.
1, 2, 3

Part 13 **For each problem, make a number family and answer the question.**

a. 56 of the animals in a kennel were cats. The kennel had a total of 402 animals. How many animals were not cats?

b. A large building had 400 steel doors and 266 doors that were not steel. How many doors did the building have in all?

c. Bob is 14 pounds lighter than his sister Milly. If Bob weighs 87 pounds, how much does Milly weigh?

d. Richard has $95 more than Sarah in his savings account. If Richard has $342, how much does Sarah have saved?

Part J

	green	red	total
experienced	48		
inexperienced		70	101
workers		121	

	green	red	total
experienced	48	51	99
inexperienced	31	70	101
workers	79	121	200

Questions
a. the red crew
b. 79 workers
c. 99 workers
d. 200 workers
e. 31 workers

There are 10 marbles for every 2 bags.
If there are 3 bags, how many marbles are there?
Marbles
Bags
Lin really works fast.
Yeah, but I think she lost her marbles.

Lesson 46

Part 1

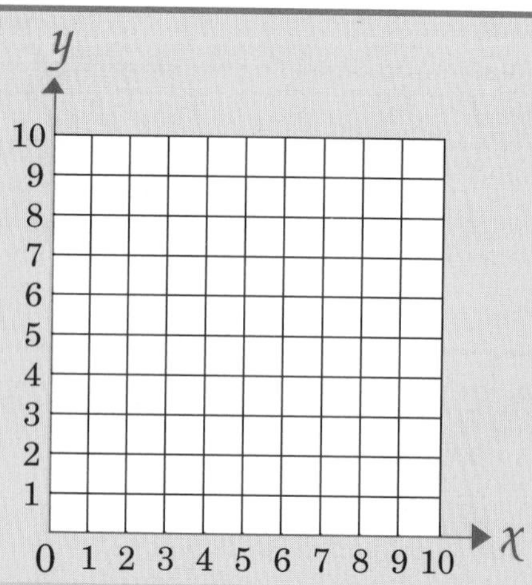

- This kind of grid is a coordinate system.
- The coordinate system has an arrow for the x direction and an arrow for the y direction.

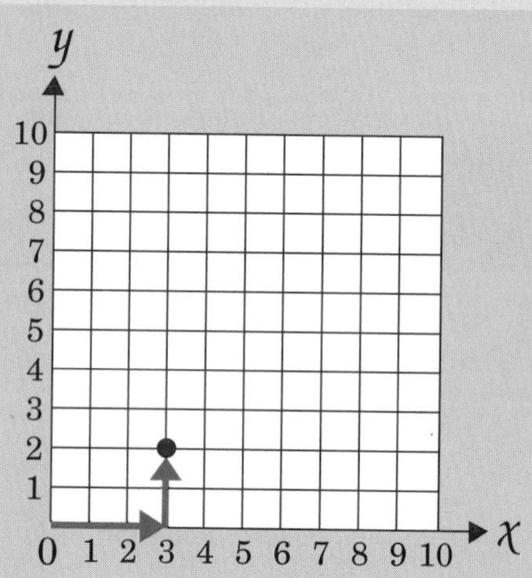

- You can tell about any point on the coordinate system by telling first about x and then about y.

$x = 3$, $y = 2$

- The point is 3 places in the x direction: →

  and 2 places in the y direction: ↑

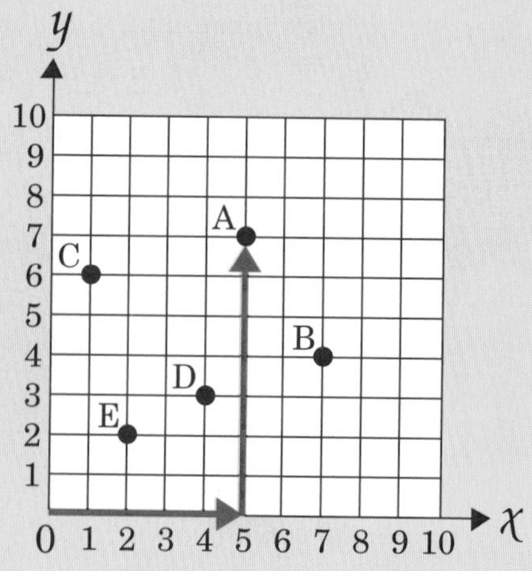

- Here are the x and y values for the point at A: $x = 5$, $y = 7$
- To get to point A, you go 5 places in the x direction, then 7 places up for y.

Part 2

For each item, make a number family that has three names and a fraction for one of the names.

> ***Sample***
> The weight of the boat was $\frac{5}{4}$ the weight of the trailer.

a. Car M went $\frac{4}{5}$ the speed of car T.

b. Car T went $\frac{4}{3}$ the speed of car Z.

c. Harry's age was $\frac{9}{10}$ the age of his dog.

d. The moose ate $\frac{7}{5}$ the amount the horse ate.

e. The boat went $\frac{9}{5}$ the distance the barge went.

Part 3

Make a table. Fill in all the missing numbers. Answer each question.

> There were ducks and geese on two ponds. The ponds were Hill Pond and Round Pond.

Facts

1. There were 28 geese on Round Pond.
2. There were 45 ducks on Hill Pond.
3. The were 70 ducks on Round Pond.
4. The total number of geese on both ponds was 45.

Items

a. On Hill Pond, there were 35 more bugs than geese. How many bugs were there on Hill Pond?

b. On Round Pond, there were 22 fewer boats than ducks. How many boats were on Round Pond?

c. On Hill Pond, there were 25 fewer loons than ducks. How many loons were on Hill Pond?

Part 4 **For each item, write an equation that shows only prime factors.**

a. 18 = 6 x 3

b. 24 = 4 x 6

c. 63 = 9 x 7

d. 16 = 4 x 4

Part 5 **Copy the table. Fill in the missing values.**

	Decimal	Fraction
a.	.73	
b.		$\frac{19}{1000}$
c.	.7	
d.	.25	
e.		$\frac{4}{100}$
f.		$\frac{305}{1000}$

Independent Work

Part 6 **For each problem, make a ratio table and answer the questions.**

a. In Rover City, the ratio of bald men to all men was 3 to 13. There were 600 bald men in Rover City.

1. How many total men were in Rover City?
2. How many men were not bald?

b. Tom made a mixture of vitamins and fish food. Tom used 3 ounces of vitamins for every 11 ounces of fish food. Tom fixed a mixture that weighed 490 ounces.

1. How many ounces of vitamins were in the mixture?
2. How many ounces of fish food were in the mixture?

Part 7 **Copy each problem and work it.**

a. $51\overline{)476}$

b. $92\overline{)279}$

c. $17\overline{)145}$

Part 8 **Copy and complete each equation.**

a. $\frac{4}{3} = \frac{20}{\blacksquare}$

b. $\frac{\blacksquare}{9} = 7$

c. $8 = \frac{\blacksquare}{5}$

d. $\frac{7}{9} = \frac{\blacksquare}{36}$

e. $\frac{\blacksquare}{3} = 9$

f. $\frac{4}{3} \times 5 \times \frac{1}{2} = \blacksquare$

g. $\frac{8}{3} - \frac{2}{3} - \frac{6}{3} = \blacksquare$

h. $\frac{82}{3} = \blacksquare$

Part 9 **Make a fraction number family. Box the answer to each question.**

a. $\frac{3}{11}$ of the cups were not green. What fraction of the cups were green?

b. 2 of the sweaters were dirty. The rest were clean. There were 9 sweaters. What fraction of the sweaters were clean?

c. 7 pounds of meat was cooked. There were 8 pounds of meat. What fraction of the meat was not cooked?

Part 10 **For each problem, make a number family and answer the question.**

a. The factory employs 285 more people than the mill. The factory employs 1420 people. How many people are employed at the mill?

b. Billy is 59 years younger than his grandfather. Billy is 13 years old. How old is his grandfather?

c. This week Millie scored 35 points. The rest of her team scored 57 points. How many points did her team score in all?

Part 11 **Write the complete equation for each lettered angle.**

a.

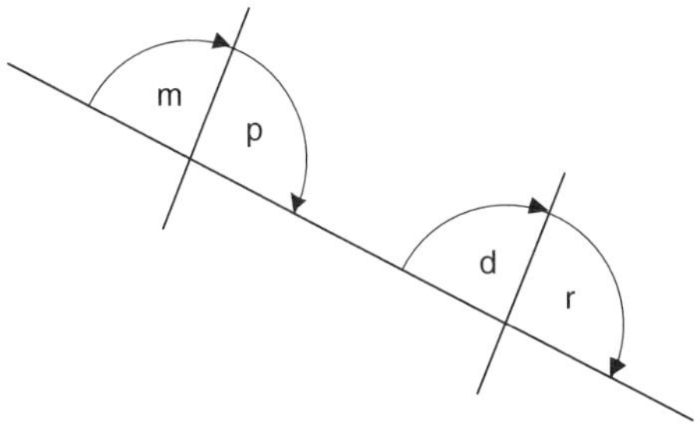

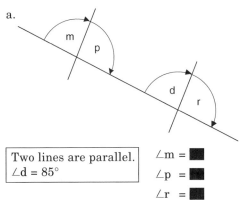

Two lines are parallel.
$\angle d = 85°$

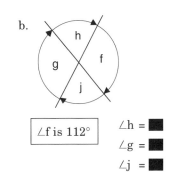

$\angle m$ = ■

$\angle p$ = ■

$\angle r$ = ■

b.

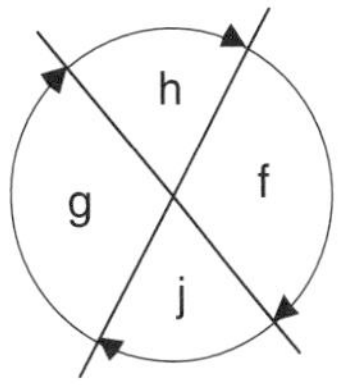

$\angle f$ is 112°

$\angle h$ = ■

$\angle g$ = ■

$\angle j$ = ■

Lesson 47

Part 1

- A fraction is **simplified** if it has the smallest numerator and the smallest denominator that is possible.
- Here are the steps:

✔ You start with a fraction that is not simplified. $\frac{28}{21}$

✔ You show the **prime factors** for the numerator and the denominator. $\frac{28}{21} = \frac{2 \times 2 \times 7}{3 \times 7}$

✔ You cross out any **fractions that equal 1.** $\frac{28}{21} = \frac{2 \times 2 \times \cancel{7}}{3 \times \cancel{7}}$

✔ Then you multiply the values that are **not crossed out.**

On top, you have 2 x 2. That's 4.
On the bottom, you have 3. $\frac{28}{21} = \frac{2 \times 2 \times \cancel{7}}{3 \times \cancel{7}} = \frac{4}{3}$

So: $\frac{28}{21} = \frac{4}{3}$

Part 2

Copy each item. Cross out fractions that equal 1. Write the simplified fraction. Below, write the simple equation.

a. $\frac{15}{40} = \frac{3 \times 5}{2 \times 2 \times 2 \times 5} =$ [redacted]

[redacted] = [redacted]

b. $\frac{10}{12} = \frac{2 \times 5}{2 \times 2 \times 3} =$ [redacted]

[redacted] = [redacted]

c. $\frac{18}{12} = \frac{2 \times 3 \times 3}{2 \times 2 \times 3} =$ [redacted]

[redacted] = [redacted]

Part 3 **For each sentence, make a number family with three names and a fraction.**

a. The distance to the mountain was $\frac{5}{8}$ the distance to the park.

b. The tree was $\frac{4}{9}$ as old as the bush.

c. The weight of the turtle was $\frac{11}{7}$ the weight of the rock.

d. Jan ate $\frac{5}{3}$ the amount that Doris ate.

Part 4 **Copy each problem and write the missing fraction.**

a. $25(\blacksquare) = 17$ b. $28(\blacksquare) = 13$ c. $30(\blacksquare) = 42$

d. $2(\blacksquare) = 10$ e. $5(\blacksquare) = 10$

Part 5 **Write the x and y values for points A through E.**

A $x = \blacksquare$, $y = \blacksquare$

B $x = \blacksquare$, $y = \blacksquare$

C $x = \blacksquare$, $y = \blacksquare$

D $x = \blacksquare$, $y = \blacksquare$

E $x = \blacksquare$, $y = \blacksquare$

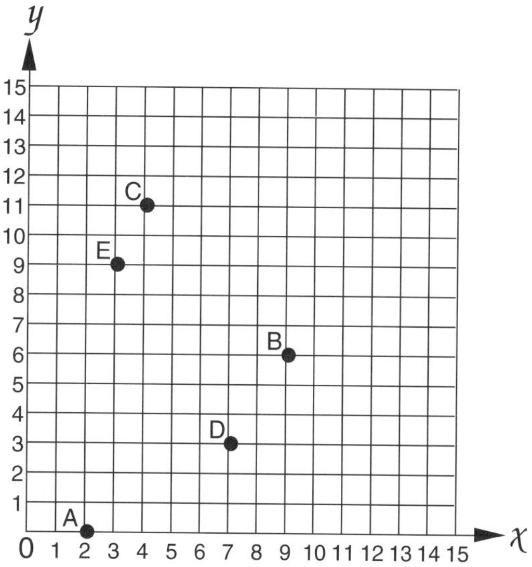

Part 6 Make a table. Fill in all the missing numbers. Answer each question.

Men and women work at two office buildings—the Billings Building and Hayward House.

Facts

1. 175 women work at Hayward House.
2. 110 men work at Hayward House.
3. 82 men work in the Billings Building.
4. The total number of employees in the Billings Building is 209.

Items

a. There are 65 fewer desks than total employees in the Billings Building. How many desks are in the Billings Building?

b. There are 256 more windows than women at Hayward House. How many windows are in Hayward House?

c. The total number of men working at both buildings is 58 less than the number of phones in both buildings. How many phones are in both buildings?

Part 7 Write the decimal number for each item.

a. $\frac{305}{100}$ b. $\frac{246}{10}$ c. $\frac{36}{100}$ d. $\frac{36}{1000}$ e. $\frac{36}{10}$

Independent Work

Part 8 For each problem, make a fraction number family and a ratio table. Answer the questions.

a. $\frac{2}{7}$ of the workers rode the bus to work. There were 420 workers.

1. How many did not ride the bus?
2. How many workers rode the bus?

b. In a deck of playing cards, $\frac{9}{13}$ of the cards are number cards. There are 52 cards in a deck.

1. How many are number cards?
2. How many are cards that do not have a number?

Part 9 **Use an estimate to check the answer to each problem. Copy each problem that is wrong and rework it.**

a. $\begin{array}{r} 1847 \\ +\ 1635 \\ \hline 2212 \end{array}$ b. $\begin{array}{r} 65 \\ -\ 27 \\ \hline 92 \end{array}$ c. $\begin{array}{r} 458 \\ +\ 323 \\ \hline 781 \end{array}$

d. $\begin{array}{r} 5076 \\ +\ 2324 \\ \hline 2752 \end{array}$ e. $\begin{array}{r} 392 \\ -\ 286 \\ \hline 106 \end{array}$

Part 10 **Work each ratio problem and answer the question.**

a. Joe ran 16 yards every 5 seconds. How many yards did Joe run if he ran at the same speed for 30 seconds?

b. A farmer can plow 4 acres on 5 gallons of fuel. How many gallons of fuel does the farmer need to plow 72 acres?

c. There are sheep and cows on a ranch. Sheep ate 16 bushels of feed for every 9 bushels cows ate. How many bushels of feed did sheep eat for every 18 bushels the cows ate?

Part 11 **Find the area and perimeter of each figure.**

a.
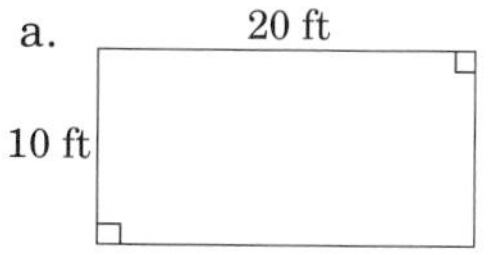

20 ft
10 ft

b.
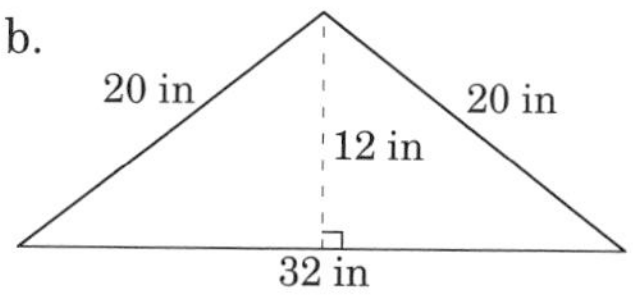

20 in
20 in
12 in
32 in

c.
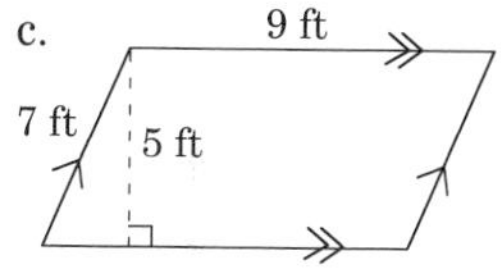

9 ft
7 ft
5 ft

Part 12 **Write an equation to show the prime factors for each number.**

a. 36 = [redacted]

b. 35 = [redacted]

c. 50 = [redacted]

Part 13 **Write the complete equation for each lettered angle.**

∠c = 43°

∠d = [redacted]

∠k = [redacted]

∠m = [redacted]

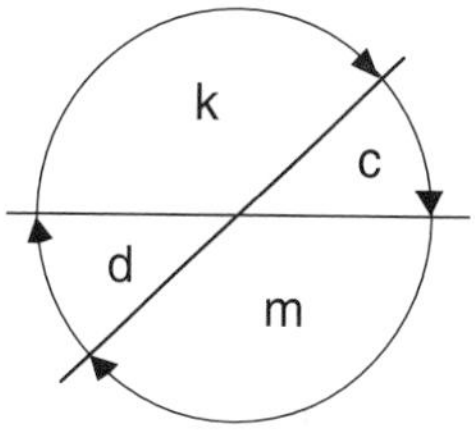

k
c
d
m

I got winded after running only ten yards. I ran through Jamie's yard, and Andy's yard, and Ginger's yard ...
Not those kind of yards.

Lesson 48

Part 1

Copy each item. Cross out fractions that equal 1. Write the simplified fraction. Below, write the simple equation.

a. $\frac{10}{30} = \frac{2 \times 5}{2 \times 3 \times 5} = \blacksquare$

$\blacksquare = \blacksquare$

b. $\frac{4}{8} = \frac{2 \times 2}{2 \times 2 \times 2} = \blacksquare$

$\blacksquare = \blacksquare$

c. $\frac{12}{18} = \frac{2 \times 2 \times 3}{2 \times 3 \times 3} = \blacksquare$

$\blacksquare = \blacksquare$

d. $\frac{28}{60} = \frac{2 \times 2 \times 7}{2 \times 2 \times 3 \times 5} = \blacksquare$

$\blacksquare = \blacksquare$

Part 2

- You've worked with sentences that use a fraction to compare two things.
- Here's a rule about these comparisons:

The thing that something is compared to equals 1.

- If something is compared to an elephant, the elephant is 1.
- If something is compared to a book, the book is 1.
- Here's a sentence:

Doris weighed $\frac{3}{4}$ as much as Edna.

- The sentence compares somebody to Edna. So Edna is 1.
- Here are some more sentences:

Doris ran $\frac{6}{5}$ the distance that Tony ran.

The boat costs $\frac{3}{5}$ as much as the car.

Part 3 **For each sentence, make a number family with all the names and the fractions.**

a. There were $\frac{7}{5}$ as many oranges as apples.

b. The weight of the apples was $\frac{4}{3}$ the weight of the turnips.

c. Amy is $\frac{3}{7}$ the age of her sister.

d. Bike C went $\frac{5}{4}$ the speed of bike F.

e. The length of the boat was $\frac{4}{9}$ the length of the pier.

Part 4 **Copy each problem and write the missing fraction.**

a. 25(■) = 8 b. 60(■) = 13 c. 52(■) = 45

d. 8(■) = 72 e. 4(■) = 20

Part 5 **Copy the table. Write the decimal number that equals each fraction.**

	Fraction	Decimal
a.	$\frac{103}{100}$	
b.	$\frac{19}{1000}$	
c.	$\frac{7}{10}$	
d.	$\frac{425}{10}$	
e.	$\frac{4}{100}$	
f.	$\frac{305}{100}$	

Part 6 **Copy and complete the table.**

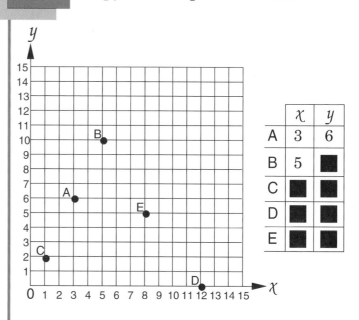

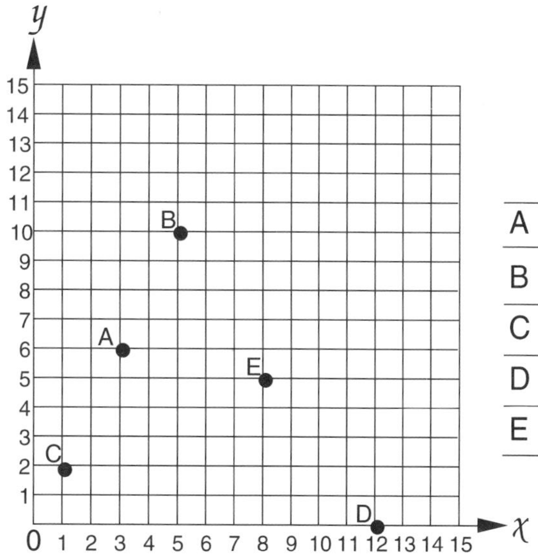

	x	y
A	3	6
B	5	■
C	■	■
D	■	■
E	■	■

Part 7 **Make a table and fill in all the missing numbers. Then answer each question.**

There are walnuts and hazelnuts in Hendricks Park and Meadow Park.

Facts

1. In Meadow Park, there is a total of 357 nuts.
2. There are 118 walnuts in Hendricks Park.
3. There are 266 walnuts in Meadow Park.
4. There are 472 hazelnuts in Hendricks Park.

Items

a. In Meadow Park, there are 307 fewer hazelnuts than honeybees. How many honeybees are in Meadow Park?

b. In Hendricks Park, there are 170 more nuts than flowers. How many flowers are in Hendricks Park?

Independent Work

Part 8 **For each problem, make a fraction number family and a ratio table. Answer the questions.**

a. $\frac{2}{3}$ of the dogs wore collars. There were 66 dogs.

1. How many wore collars?
2. How many did not wear collars?

b. $\frac{2}{17}$ of the animals in the pond were worms. In the pond, 450 animals were not worms.

1. How many worms were in the pond?
2. How many total animals were in the pond?

Part 9 **Copy and work each problem.**

a. $53\overline{)241}$

b. $36\overline{)298}$

c. $42\overline{)331}$

Part 10 **For each item, make a table and complete it.**

a. Divide each number by 2. Then subtract 10.
136, 200, 20

b. Add 5 to each number. Then multiply the sum by 12.
1, 4, 0

Part 11

For each problem, make a number family and answer the question.

a. $\frac{1}{9}$ of the plants are not green. What fraction of the plants are green?

b. Bryan saved $210 more than Debbi. Debbi saved $843. How much did Bryan save?

c. 3 of the children were boys. There were 8 girls. What fraction of the children were girls?

d. There were 210 fewer students at the elementary school than at the high school. There were 498 students at the elementary school. How many students were at the high school?

e. It is 158 miles further to Clear Lake than to Muddy Lake. It is 223 miles to Clear Lake. How far is it to Muddy Lake?

f. I do not work for $\frac{2}{7}$ of the days. For what fraction of the days do I work?

Part 12

Work each problem.

a. $\frac{9}{15} - \frac{8}{15} + \frac{1}{15} = \blacksquare$

b. $9\overline{)64}$

c. $3 \times \frac{4}{5} = \blacksquare$

d. $4\overline{)175}$

e. $\frac{4}{5} = \frac{\blacksquare}{60}$

f. $\frac{3}{8} = \frac{27}{\blacksquare}$

g. $5 = \frac{\blacksquare}{11}$

Part 13

Write the equation to show the prime factors for each number.

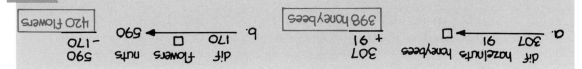

a. 72 b. 33 c. 32 d. 50

Part J

a. dif hazelnuts honeybees
307 91 → □
307 + 91 = 398 honeybees

b. dif flowers nuts
170 □ ← 590
590 − 170 = 420 flowers

Lesson 49

Part 1 For each statement, make a number family with three numbers.

This table shows the number of cars and trucks on two streets.

a. There were 15 more cars than trucks on J Street.

b. There were 41 more trucks than cars on D Street.

c. There were 48 fewer vehicles on D Street than on J Street.

	cars	trucks	vehicles
D Street	8	49	57
J Street	60	45	105
total	68	94	162

Part 2 For each statement, make a number family with three numbers.

This table shows the number of red butterflies and not-red butterflies spotted in 2 months.

a. In August, there were 24 fewer red butterflies than butterflies that were not red.

b. In September, there were 12 more red butterflies than butterflies that were not red.

c. The total number of butterflies for August was 400 less than the total for September.

	August	September	total
red	292	510	802
not red	316	498	814
butterflies	608	1008	1616

Part 3 **Copy each fraction. Show the prime factors for the numerator and the denominator. Write the simplified fraction.**

Sample fraction $\frac{8}{18}$

a. $\frac{15}{25}$ b. $\frac{10}{40}$ c. $\frac{42}{35}$ d. $\frac{8}{16}$

Part 4 **Make a fraction number family for each sentence.**

a. The sale price of the couch is $\frac{4}{7}$ the regular price of the couch.

b. The weight of the flour is $\frac{8}{3}$ the weight of the potatoes.

c. The number of people at the ball game was $\frac{5}{9}$ the number of people at the concert.

d. Factory A makes $\frac{7}{4}$ the quantity that Factory B makes.

Part 5 **Copy the table. Write the fraction that equals each decimal number.**

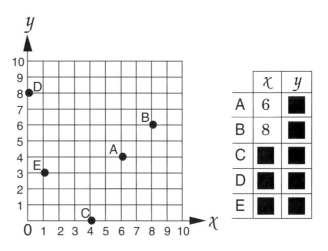

	Fraction	Decimal
a.		3.04
b.		1.705
c.		1.24
d.		17.6
e.		.03

Part 6 **Copy and complete the table.**

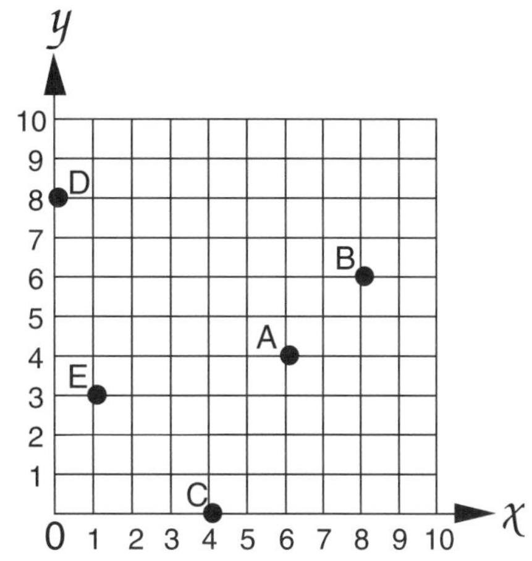

	x	y
A	6	■
B	8	■
C	■	■
D	■	■
E	■	■

Part 7 Copy each problem. Below, write the complete equation to show what f equals.

Sample 1	$6f = 48$ $f = \frac{48}{6} = 8$	***Sample 2***	$4f = 188$ $f = \frac{188}{4} = ■$

a. $5f = 936$ b. $4f = 338$ c. $9f = 364$ d. $8f = 450$

Independent Work

Part 8 For each problem, write the ratio equation. Answer the question.

a. In a pond, the ratio of all fish to older fish is 7 to 3. If there are 630 fish in the pond, how many are older fish?

b. A shirt maker uses 4 feet of thread for every 8 buttons. The shirt maker used 736 feet of thread. How many buttons did the shirt maker use?

c. A worker painted 7 boards every 6 minutes. The worker painted 910 boards. How many minutes did this work take?

Part 9 Write an equation to show the prime factors for each number.

a. 4 b. 20 c. 100 d. 120

Part 10 Copy and complete each table.

a.

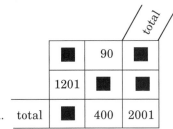

			total
	■	90	■
	1201	■	■
total	■	400	2001

b.

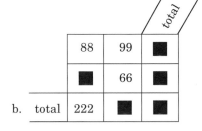

			total
	88	99	■
	■	66	■
total	222	■	■

Part 11 Copy and complete the table.

	Division	Fraction equation	Multiplication
a.	$6\overline{)431}$	■ = ■	■(■) = ■
b.	$6\overline{)120}$		
c.	$4\overline{)902}$		

Part 12 Write parallel, not parallel, or intersecting for each group of lines.

a. b. c. d.

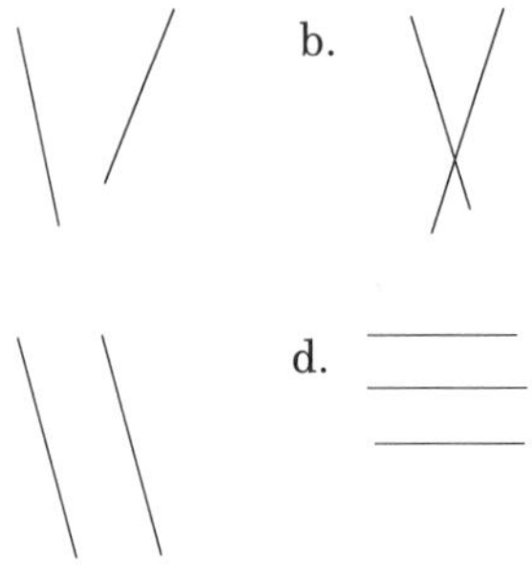

Part 13 **For each item, make a number family and answer the question.**

a. The yard was 280 feet longer than the house. The yard was 335 feet long. How long was the house?

b. The busboy earned $118 less than the cook each week. The busboy earned $127 each week. How much did the cook earn?

c. Briskhorn Zoo had 7124 animals. Elkhorn Zoo had 5811 animals. How many fewer animals did Elkhorn Zoo have than Briskhorn Zoo?

d. 4780 of the animals at Briskhorn Zoo were mammals. There were 7124 animals in Briskhorn Zoo. How many animals were not mammals?

Part 14 **Make a fraction number family for each item. Box the fraction that answers each question.**

a. $\frac{2}{3}$ of the participants were children. What fraction were adults?

b. $\frac{2}{10}$ of the questions were difficult. What fraction were not difficult?

Part 15 **Copy and complete each equation.**

a. $11(\blacksquare) = 5$ b. $57(\blacksquare) = 19$ c. $8(\blacksquare) = 21$ d. $16(\blacksquare) = 4$

Lesson 50

Part 1 **Copy each problem. Below, write the complete equation.**

a. $5f = 384$
$f = ■ = ■$

b. $3f = 384$
$f = ■ = ■$

c. $6f = 940$
$f = ■ = ■$

d. $2f = 620$
$f = ■ = ■$

Part 2 **For each item, write an equation to show the decimal value and the fraction it equals.**

a. 5.03 b. 42.8 c. .15 d. 1.5

Part 3 **Check the x and y values for each point. If any row in the table is wrong, rewrite the row with the correct x and y values.**

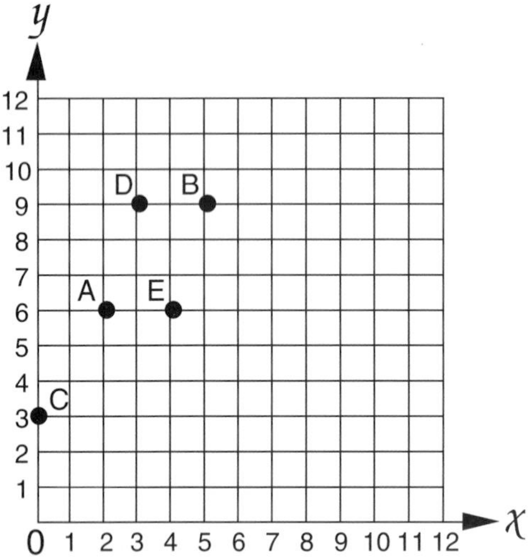

	x	y
A	3	6
B	5	9
C	0	4
D	9	3
E	4	6

Test 5

Part 1 **Copy and work each problem. The estimation answer may be too large or too small.**

a. $24\overline{)134}$ b. $71\overline{)498}$ c. $17\overline{)160}$

Part 2 **For each problem, make a fraction number family and a ratio table. Answer the questions.**

a. In a grove of trees, there were broadleaf trees and trees with needles. $\frac{2}{7}$ of the trees were broadleaf trees. The grove had a total of 700 trees.

1. How many trees were broadleaf trees?
2. How many were trees with needles?

b. $\frac{8}{11}$ of the material used for each window was glass. The rest was wood. A factory used 232 square feet of glass to make windows.

1. How many square feet of wood did the factory use?
2. How many square feet of material was used in all?

Part 3 **Copy each problem. Write the missing fraction.**

a. $17(\blacksquare) = 15$ b. $9(\blacksquare) = 19$

c. $3(\blacksquare) = 11$ d. $46(\blacksquare) = 39$

Part 4

For each problem, make a fraction number family and box the fraction that answers the question.

a. There were 23 people in a park. 17 of them were children. What fraction of the people were not children?

b. Richard mixed paint with water. The mixture he made weighed 11 pounds. 4 pounds of the mixture was water. What fraction of the mixture was paint?

c. $\frac{2}{13}$ of the people on the bus wore a hat. What fraction of the people did not wear a hat?

Part 5

Copy and work each problem.

a. $\frac{3}{5} \times 2 \times \frac{1}{8} = \blacksquare$

b. $1 \times \frac{4}{7} \times \frac{7}{4} = \blacksquare$

Part 6

Write the equation to show the prime factors for each number.

a. 22

b. 40

c. 72

Lesson 51

Part 1

- You can write decimal numbers that are more than 1 as mixed numbers. You just read the decimal number. That tells you how to write the mixed number.

$$3.14 = 3\frac{14}{100}$$

$$6.2 = 6\frac{2}{10}$$

Part 2

For each item, write the complete equation with the decimal number and the mixed number it equals.

a. 1.07 b. 14.002 c. 3.5 d. 216.23

Part 3

Copy and complete the table.

	Decimal	Fraction
a.	8.5	
b.	11.09	
c.	6.35	
d.	.8	
e.	25.1	
f.	.047	

Part 4

For each sentence, make a number family.

a. $\frac{3}{5}$ of the red chairs were old.

b. The red chairs were $\frac{3}{4}$ the age of the blue chairs.

c. The dog's weight was $\frac{5}{3}$ the cat's weight.

d. $\frac{3}{7}$ of the apples were ripe.

e. Mary was $\frac{2}{5}$ as old as Sarah.

f. $\frac{1}{5}$ of the cars were new.

Part 5 Copy each problem. Below, write the complete equation to show what f equals.

a. $7f = 497$ $f = ■ = ■$

b. $3f = 762$ $f = ■ = ■$

c. $4f = 762$ $f = ■ = ■$

d. $2f = 762$ $f = ■ = ■$

Part 6 For each statement, make a number family. Answer the questions.

> This table shows the number of guests staying at two hotels on Friday and Sunday.

Statement 1

At hotel A, the number of guests on Friday was 211 less than the number of guests on Sunday.

Statement 2

On Sunday, 645 more guests were in hotel B than in hotel A.

	hotel A	hotel B	total
Friday	■	1243	■
Sunday	672	■	■
total	■	■	■

Items

a. How many guests stayed at hotel B on Sunday?

b. At hotel B, what was the total number of guests for both days?

c. On Sunday, were more guests staying at hotel A or hotel B?

d. What was the total number of guests for both hotels on both days?

e. On which day did fewer guests stay at hotel A?

Part 7 Simplify each fraction.

a. $\frac{28}{24}$ b. $\frac{35}{10}$ c. $\frac{5}{50}$

Independent Work

Part 8 **For each problem, make a ratio table. Answer the questions the problem asks.**

a. In a coin shop, the ratio of gold coins to silver coins is 5 to 7. There are 2400 coins in the shop.
   1. How many are gold coins?
   2. How many are silver coins?

b. The ratio of dead trees to living trees in a forest is 2 to 15. There are 30 dead trees in the forest.
   1. How many total trees are in the forest?
   2. How many living trees are in the forest?

Part 9 **Copy and work each problem.**

a. $51\overline{)236}$

b. $24\overline{)236}$

c. $93\overline{)236}$

Part 10 **For each problem, make the fraction number family. Box the answers to the questions.**

a. $\frac{2}{3}$ of the apples were rotten. What's the fraction for all the apples?

b. $\frac{4}{79}$ of the cars had poor tires. What fraction of the cars had good tires?

c. $\frac{1}{6}$ of the pie had been eaten. What fraction of the pie remains?

Part 11 **Work each problem. Write each answer as a fraction.**

a. $1 - \frac{3}{8} = \blacksquare$

b. $\frac{3}{5} + 2 = \blacksquare$

c. $\frac{3}{5} + \frac{2}{5} = \blacksquare$

d. $\frac{1}{8} + \frac{3}{8} = \blacksquare$

e. $\frac{8}{4} + \frac{3}{4} + \frac{1}{4} = \blacksquare$

f. $14(\blacksquare) = 35$

g. $74(\blacksquare) = 1$

Part 12 **Find the perimeter and the area of each figure.**

a.

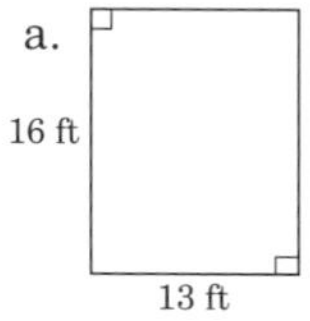

b.

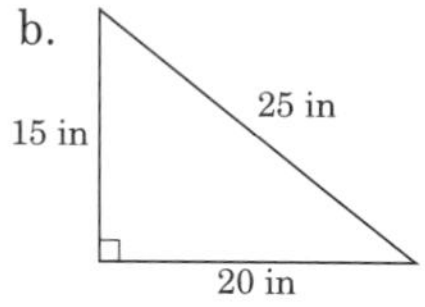

Part 13 **For each item, write an equation to show the fraction and the decimal value it equals.**

a. $\frac{36}{1000}$

b. $\frac{3}{100}$

c. $\frac{120}{100}$

d. $\frac{405}{10}$

Part 14 **Copy the table. For each row, write the multiplication equation, the division problem and the answer, and the fraction equation.**

	Multiplication	Division	Fraction equation
a.	5 (■) = 120		
b.		$8\overline{)584}$	
c.			$\frac{128}{8}$ = ■

Part J

	Decimal	Fraction
a.	8.5	$\frac{85}{10}$
b.	11.09	$\frac{1109}{100}$
c.	6.35	$\frac{635}{100}$
d.	.8	$\frac{8}{10}$
e.	25.1	$\frac{251}{10}$
f.	.047	$\frac{47}{1000}$

Part K

	hotel A	hotel B	total
Friday	461	1243	1704
Sunday	672	1317	1989
total	1133	2560	3693

I figured out what fraction of the trees are not large. It's the top $\frac{1}{3}$.
I don't think that's what they wanted you to figure out.
$\frac{6}{7}$ of the trees are large. What's the fraction for the trees that are not large?

Lesson 52

Part 1

Decimal values work like fractions.

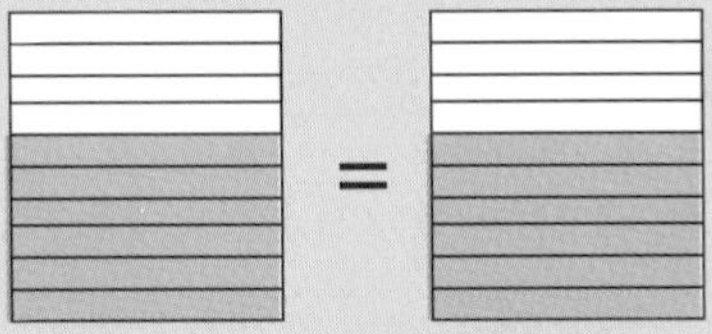

$$.6 = \frac{6}{10}$$

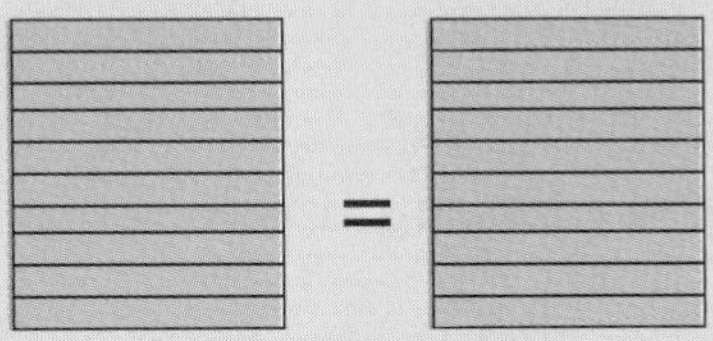

$$\frac{10}{10} = 1.0 = 1$$

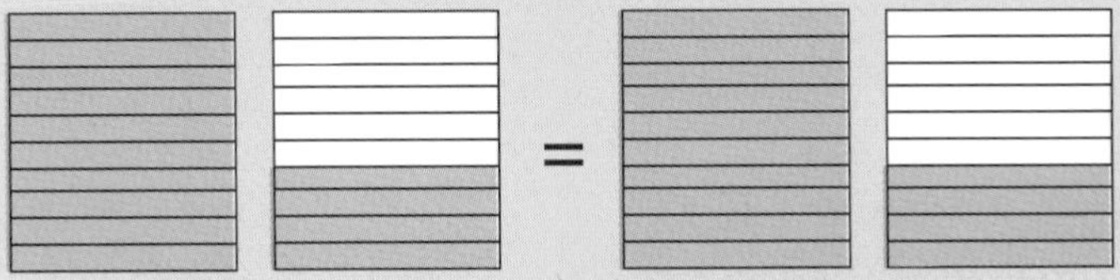

$$\frac{14}{10} = 1.4 = 1\frac{4}{10}$$

Part 2

For each item, write the equation that shows the fraction, the decimal number and the mixed number it equals.

Sample $\frac{15}{10}$

a. $\frac{99}{10}$ b. $\frac{541}{100}$ c. $\frac{126}{10}$ d. $\frac{409}{100}$

Part 3

Copy each problem. Write the complete equation to show the fraction and the whole number or mixed number that f equals.

a. $5f = 266$ b. $7f = 266$ c. $2f = 49$ d. $3f = 996$

Part 4 Make a fraction number family for each sentence.

a. 5 out of every 8 cards are blue.

b. The cat weighed $\frac{3}{8}$ as much as the dog.

c. The sale price of the TV was $\frac{3}{7}$ the regular price.

d. The barn was $\frac{1}{5}$ the size of the pasture.

e. $\frac{4}{5}$ of the items in the store were on sale.

f. The height of the tree was $\frac{9}{8}$ the height of the pole.

Part 5 Figure out all the missing numbers in the table.

This table shows the number of 9th graders and 10th graders who attend two high schools.

Statement 1

At Jefferson High, there are 15 fewer 10th graders than 9th graders.

Statement 2

There are 27 more 9th graders at Madison High than there are at Jefferson High.

	9th graders	10th graders	total
Jefferson High	228	■	■
Madison High	■	284	■
total	■	■	■

Part 6 Copy each fraction. Show the prime factors for the numerator and the denominator. Write the simplified fraction.

a. $\frac{10}{25}$ b. $\frac{4}{12}$ c. $\frac{12}{9}$ d. $\frac{24}{30}$

Part 7

- When you work division problems on your calculator, the calculator shows mixed numbers as **decimal values.**

  [5][0][÷][4][0][=][1.25]
  That's $1\frac{25}{100}$.

  [2][0][÷][8][=][2.5]
  That's $2\frac{5}{10}$.

- For some division problems, the calculator gives a very long answer. When you divide 10 by 3, you get this answer: 3.33333333. That's the decimal value for $3\frac{1}{3}$.

  [1][0][÷][3][=][3.33333333]
  That's $3\frac{1}{3}$.

Part 8

Work each problem on your calculator. Write the answer as a mixed number.

a. [5][÷][2][=]   b. [1][7][÷][8][=]   c. [1][3][÷][4][=]   d. [1][6][÷][5][=]

Independent Work

Part 9

For each problem, make a fraction number family. Box the answers to the questions.

a. $\frac{18}{37}$ of the fans cheered for the home team. What fraction of the fans cheered for the visiting team?

b. 5 out of every 9 apples were red. What fraction of the apples were red?

Part 10

For each problem, make a ratio table. Answer each question.

a. 4 out of every 7 animals on Dean's farm are pigs. The rest are not pigs. There are 140 animals on Dean's farm.
   1. How many animals are pigs?
   2. How many animals are not pigs?

b. In a pond, there are 8 green frogs for every 3 frogs that are not green. There are 60 frogs that are not green.
   1. How many green frogs are in the pond?
   2. How many total frogs are in the pond?

Part 11

Copy and complete each equation.

$\angle d = 135°$
$\angle a =$ ■
$\angle b =$ ■

$\angle g =$ ■
$\angle f =$ ■

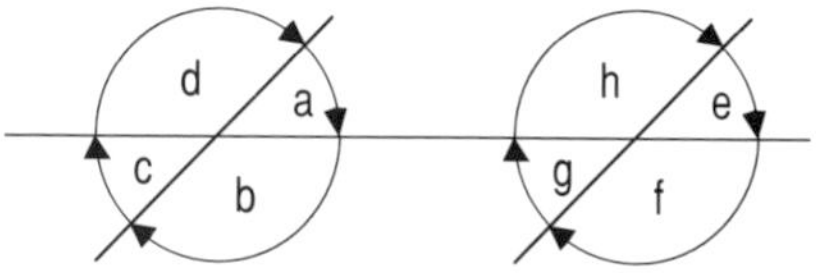

Part 12 **For each row, write the decimal value and the fraction it equals.**

	Decimal	Fraction
a.	63.1	
b.		$\frac{74}{1000}$
c.		$\frac{418}{100}$
d.	.054	
e.		$\frac{92}{10}$

Part 13 **Copy and complete the table.**

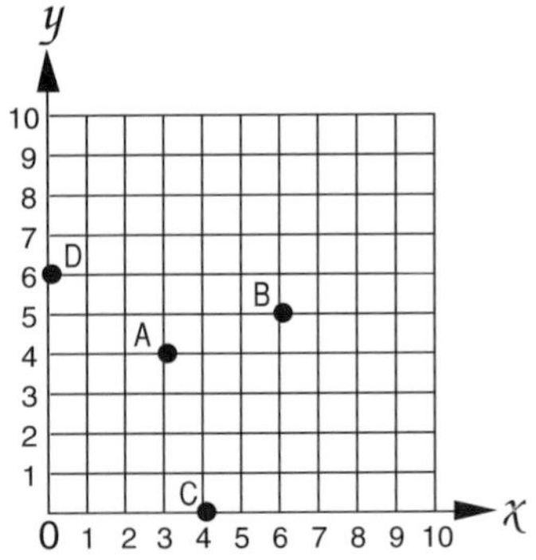

	x	y
A	■	■
B	■	■
C	■	■
D	■	■

Part 14 **Make a table and complete it.**

Multiply each number by 4. Then subtract the product from 200.

50, 30, 1, 0

Part 15 **Copy and work each problem.**

a. $91\overline{)549}$ b. $38\overline{)114}$

c. $2\overline{)5814}$ d. $53\overline{)387}$

Part 16 **Answer each question.**

This table shows the number of 9th graders and 10th graders who attend two high schools.

a. In which school are there more 9th graders?
b. How many 10th graders are in both schools?
c. What is the total number of 9th and 10th graders attending both schools?
d. What is the total number of 9th and 10th graders at Madison High?
e. In which school are there fewer 10th graders?

	9th graders	10th graders	total
Jefferson High	228 X	213 X	441
Madison High	255 X	284	539
total	483	497	980

Part J

a. $5f = 266$ $f = \frac{266}{5} = 53\frac{1}{5}$

b. $7f = 266$ $f = \frac{266}{7} = 38$

c. $2f = 49$ $f = \frac{49}{2} = 24\frac{1}{2}$

d. $3f = 996$ $f = \frac{996}{3} = 332$

Part K

a. $\frac{10}{25} = \frac{2\times\cancel{5}}{5\times\cancel{5}} = \frac{2}{5}$

b. $\frac{4}{12} = \frac{\cancel{2}\times\cancel{2}}{\cancel{2}\times\cancel{2}\times3} = \frac{1}{3}$

c. $\frac{12}{9} = \frac{2\times2\times\cancel{3}}{3\times\cancel{3}} = \frac{4}{3}$

d. $\frac{24}{30} = \frac{\cancel{2}\times2\times2\times\cancel{3}}{\cancel{2}\times\cancel{3}\times5} = \frac{4}{5}$

Lesson 53

Part 1

- You've learned that fractions equal 1 if the top value is the same as the bottom value. That's true for complicated fractions that have a **fraction** over the same **fraction.**

$$\frac{\frac{4}{8}}{\frac{4}{8}} = 1$$

- The fraction equals 1 because the value on top is the same as the value on the bottom.
- Here are some other fractions that equal 1:

$$\frac{\frac{4m}{7}}{\frac{4m}{7}} = 1 \qquad \frac{\frac{3}{85}}{\frac{3}{85}} = 1 \qquad \frac{\frac{71}{2}}{\frac{71}{2}} = 1$$

Part 2 Copy and complete each equation.

a. $\frac{\frac{5}{8}}{\blacksquare} = 1$ b. $\frac{\blacksquare}{\frac{7}{3}} = 1$ c. $\frac{\frac{11}{5}}{\blacksquare} = 1$

Part 3 Write the equation that shows the fraction, the decimal number and the mixed number it equals.

a. $\frac{740}{100}$ b. $\frac{29}{10}$ c. $\frac{163}{10}$ d. $\frac{422}{100}$

Part 4 For each item, write the complete equation.

Sample $\frac{3}{3} = \frac{\frac{\blacksquare}{2}}{\frac{\blacksquare}{2}} = \frac{\frac{\blacksquare}{7}}{\frac{\blacksquare}{7}}$

a. $\frac{2}{2} = \frac{\frac{\blacksquare}{5}}{\frac{\blacksquare}{5}} = \frac{\frac{\blacksquare}{8}}{\frac{\blacksquare}{8}}$

b. $\frac{7}{7} = \frac{\frac{\blacksquare}{3}}{\frac{\blacksquare}{3}} = \frac{\frac{\blacksquare}{9}}{\frac{\blacksquare}{9}}$

Part 5

- Here's a difficult problem:

  The regular price of a bed is $\frac{8}{5}$ the sale price. If the regular price of the bed is \$560, what is the difference between the sale price and the regular price?

- The first sentence tells how to make a fraction number family.
- The big number is the **regular price.**

dif	sale	regular
	→	$\frac{8}{5}$

- The sentence compares the regular price to the sale price. So the sale price is **1 whole.**
- The difference is $\frac{3}{5}$.

dif	sale	regular
$\frac{3}{5}$	$\frac{5}{5}$ →	$\frac{8}{5}$

- You know that the numerators are ratio numbers. You put them in a ratio table. Then you put in the number the problem gives for the regular price of the bed.

dif	3	
sale	5	
regular	8	560

- Now you can figure out the sale price of the bed and how much you'd save if you bought the bed on sale. That's the number for difference in the second column.

Part 6

Make a number family and a ratio table for each item. Answer each question.

a. The regular price of a chair is $\frac{5}{4}$ the sale price of the chair. If you bought the chair on sale, you'd save \$12.

1. What's the regular price of the chair?
2. What's the sale price of the chair?

b. The elephant weighs $\frac{3}{8}$ as much as the rock. The elephant weighs 6000 pounds.

1. How much does the rock weigh?
2. How much more does the rock weigh than the elephant weighs?

c. The cost of the sewing machine is $\frac{7}{5}$ the cost of the TV set. The TV set costs $280.

1. How much does the sewing machine cost?
2. What's the difference between the cost of the sewing machine and the cost of the TV set?

Part 7

Work each problem with your calculator. Then write the decimal answer to each problem as a mixed number.

a. 25 ÷ 4 =

b. 117 ÷ 6 =

c. 108 ÷ 25 =

d. 100 ÷ 32 =

Part 8

Copy each fraction. Show the prime factors for the numerator and the denominator. Write the simplified fraction.

a. $\frac{21}{28}$　　b. $\frac{8}{12}$　　c. $\frac{3}{18}$　　d. $\frac{24}{30}$

Independent Work

Part 9

For each problem, make a fraction number family and answer the questions.

a. $\frac{2}{3}$ of the chairs were painted white. The rest were not. There were 46 white chairs.

1. How many chairs were there in all?
2. How many of those chairs were not white?

b. $\frac{1}{5}$ of the apples in a bag were rotten. 80 of the apples in the bag were not rotten.

1. How many apples were there in the bag?
2. How many were rotten?

Part 10

Copy each problem. Write the equation that shows the fraction and the whole number or mixed number the letter equals.

a. 8M = 17
M = ■ = ■

b. $\frac{3}{5} - \frac{2}{5} + \frac{7}{5} = R$
R = ■ = ■

c. 6J = 20
J = ■ = ■

d. $\frac{3}{5} \times 2 \times \frac{9}{3} = V$
V = ■ = ■

e. 8Q = 325
Q = ■ = ■

Part 11 Copy and work each problem.

a. $\begin{array}{r} \frac{13}{20} \\ + 1 \\ \hline ■ \end{array}$

b. $\begin{array}{r} 1 \\ - \frac{4}{5} \\ \hline ■ \end{array}$

c. $\frac{2}{3} \times 5 = ■$

d. $\frac{1}{2} \times \frac{10}{5} = ■$

e. $\frac{5}{8} - \frac{5}{8} = ■$

Part 12 For each item, write the complete equation. If the fractions shown are equivalent, write the simple equation below.

a. $\frac{3}{5}, \frac{231}{375}$

b. $\frac{2}{3}, \frac{66}{99}$

c. $\frac{5}{7}, \frac{15}{21}$

d. $\frac{7}{4}, \frac{28}{12}$

Part 13 Make a table and complete it.

Divide each number by 2.
Then add 5.

10, 8, 20, 2, 0

Part 14 Copy and complete the table.

	Multiplication	Division
a.	24 (■) = 216	
b.	12 (■) = 100	
c.		$8\overline{)125}$

Part J

a. $\frac{21}{28} = \frac{3 \times 7}{2 \times 2 \times 7} = \frac{3}{4}$ b. $\frac{8}{12} = \frac{2 \times 2 \times 2}{2 \times 2 \times 3} = \frac{2}{3}$ c. $\frac{3}{18} = \frac{3}{2 \times 3 \times 3} = \frac{1}{6}$ d. $\frac{24}{30} = \frac{2 \times 2 \times 2 \times 3}{2 \times 3 \times 5} = \frac{4}{5}$

Lesson 54

Part 1 **Copy the table. Fill in the missing values.**

	Fraction	Decimal	Mixed number
a.	$\frac{302}{100}$		
b.			$5\frac{8}{10}$
c.		1.015	
d.		13.4	

Part 2

- You've learned that the distance around any figure with straight sides is called the **perimeter.**

- A circle doesn't have any straight sides. So it doesn't have a perimeter. The distance around a circle is called the **circumference.**
- The first part of **circumference** has the same letters as the word **circle.**

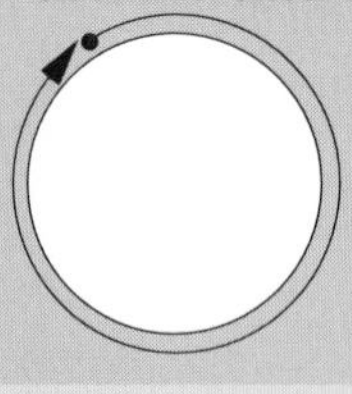

- To figure out the distance around a circle, you **can't** add up the length of each side because there are no sides. But you can **measure** the **circumference of a circle.**
- One way is to roll the circle and see how long the path is when the circle rotates one time.

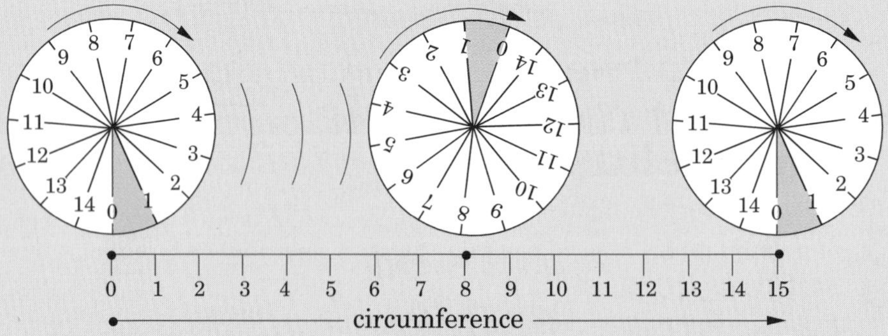

- The distance around the circle is 15 units. So the circumference is 15 units.
- The circumference of any circle is related to a line that goes through the center of that circle. That line is called the **diameter.**

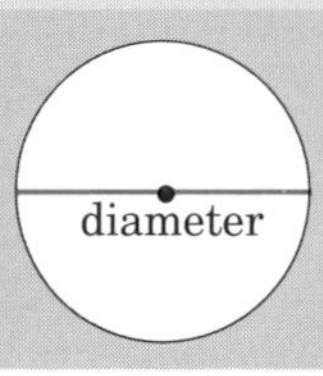

Part 3 For each circle, write the complete equation with numbers. Figure out the decimal value for each fraction.

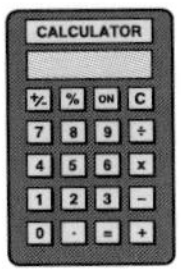

		Equation
a.	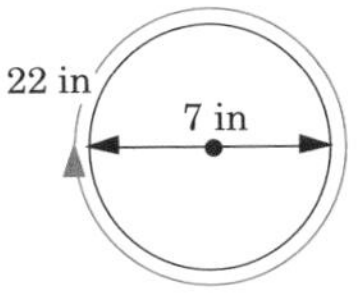	$d\left(\frac{\blacksquare}{\blacksquare}\right) = C$
b.	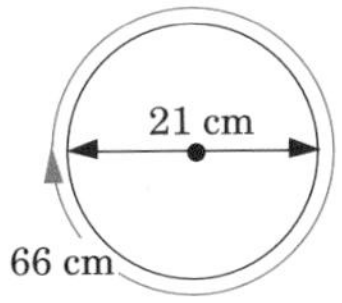	$\blacksquare\left(\frac{\blacksquare}{\blacksquare}\right) = \blacksquare$
c.	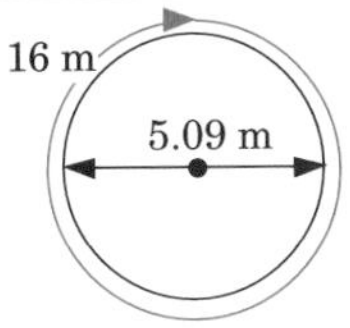	$\blacksquare\left(\frac{\blacksquare}{\blacksquare}\right) = \blacksquare$

Part 4 Copy and complete each equation.

a. $\dfrac{\blacksquare}{\frac{10}{5}} = 1$ b. $\dfrac{\frac{2}{3}}{\blacksquare} = 1$ c. $\dfrac{\frac{17}{4}}{\blacksquare} = 1$

Part 5 For each item, write the complete equation.

a. $\dfrac{5}{5} = \dfrac{\frac{\blacksquare}{3}}{\frac{\blacksquare}{3}} = \dfrac{\frac{\blacksquare}{9}}{\frac{\blacksquare}{9}}$ b. $\dfrac{6}{6} = \dfrac{\frac{\blacksquare}{5}}{\frac{\blacksquare}{5}} = \dfrac{\frac{\blacksquare}{9}}{\frac{\blacksquare}{9}}$ c. $\dfrac{2}{2} = \dfrac{\frac{\blacksquare}{10}}{\frac{\blacksquare}{10}} = \dfrac{\frac{\blacksquare}{9}}{\frac{\blacksquare}{9}}$

Part 6 Make a number family and a ratio table. Answer the questions.

a. The area of the carpet was $\frac{3}{2}$ the area of the living room. The installers had to cut 24 square meters from the carpet before it would fit in the living room.

1. What's the area of the living room?
2. What was the area of the carpet before the installers cut it?

b. The number of students on the field trip was $\frac{8}{5}$ the number of students attending the hockey practice. 35 students attended the hockey practice.

1. How many students were on the field trip?
2. How many more students were on the field trip than were at the hockey practice?

c. Lunch cost $\frac{4}{9}$ as much as dinner cost. Dinner cost $36.

1. How much did lunch cost?
2. How much less was the cost of lunch than the cost of dinner?

Part 7

- Some fractions cannot be simplified. A fraction cannot be simplified if it has no common factor in the numerator and in the denominator.

Here's: $\frac{\mathbf{14}}{\mathbf{15}} = \frac{\mathbf{2 \times 7}}{\mathbf{3 \times 5}}$

- No factor appears in **both** the numerator and denominator. The fraction cannot be simplified because you cannot cross out a fraction that equals 1.
- Here's another fraction that cannot be simplified: $\frac{\mathbf{21}}{\mathbf{10}} = \frac{\mathbf{3 \times 7}}{\mathbf{2 \times 5}}$
- No factor appears in **both** the numerator and denominator.

Part 8 Copy each fraction. Write the prime factors for the numerator and the denominator. Cross out any fractions that equal 1. Write the value you get when you multiply.

a. $\frac{8}{15}$ b. $\frac{6}{7}$ c. $\frac{14}{28}$ d. $\frac{24}{36}$

Independent Work

Part 9 Copy each equation. Write each missing value as a whole number or mixed number.

a. $\frac{48}{29} = \blacksquare$ b. $\frac{148}{52} = \blacksquare$ c. $\frac{7}{8} = \frac{\blacksquare}{72}$ d. $\frac{9}{7} = \frac{36}{\blacksquare}$

Part 10 Make a ratio table for each problem. Answer the questions.

a. The ratio of old tires to new tires at a repair shop is 5 to 9. The repair shop has a total of 168 tires.
   1. How many tires are old?
   2. How many tires are new?

b. In Elway School, 4 out of every 10 students are girls. The rest are boys. There are 540 boys at Elway School.
   1. How many girls are in the school?
   2. How many total students are in the school?

Part 11 Make a table and answer the questions.

Ace Mine and Donner's Mine sell copper and gold. The total amount of copper sold by both the mines was 496 tons. The amount of gold sold by Ace Mine was 364 tons. The amount of gold sold by Donner's Mine was 140 tons. The total amount of copper and gold sold by Ace Mine was 620 tons.

a. How many total tons of gold were sold by the two mines together?
b. What was the total amount of copper and gold sold by Donner's Mine?
c. Which mine sold more copper?
d. Which was greater, the total amount of copper sold or the total amount of gold sold?

Part 12 Write the complete equation with the missing value as a fraction.

a. $5(\blacksquare) = 4$ b. $8(\blacksquare) = 1$ c. $8(\blacksquare) = 9$ d. $15(\blacksquare) = 37$ e. $256(\blacksquare) = 8$

Part 13 Find the area and perimeter of each figure.

a.

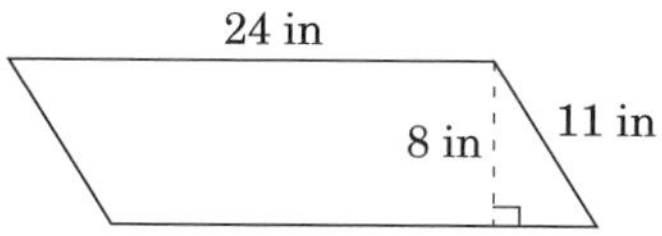

b.

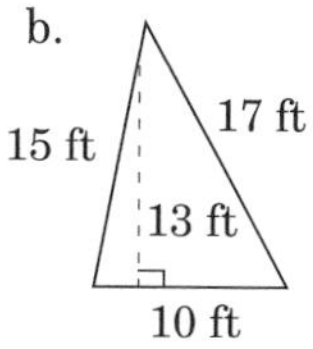

Part 14 Copy and work each problem.

a. $37\overline{)290}$

b. $56\overline{)119}$

c. $88\overline{)222}$

Lesson 55

Part 1

Here's how to round decimal numbers to the nearest hundredth:

- Look at the **third** digit after the decimal point.

  2.3851 (arrow pointing to the 5)

- If that digit is 5 or more, you do not copy the **hundredths** digit. You replace it with the digit that is 1 larger.

  2.3851
  2.39

- If the third digit after the decimal point is less than 5, you just copy the hundredths digit.

  2.3841 (arrow pointing to the 4)

  2.3841
  2.38

Part 2

Round each decimal number to the nearest hundredth.

Sample 1 7.135	*Sample 2* 7.364

a. 3.6473 b. 18.05396 c. 10.273

d. .629 e. 4.0135 f. 4.0162

Part 3

- The diameter of any circle is related to the circumference of the circle. The circumference is always **3.14 times the diameter.**

- That's true of small circles and large circles.

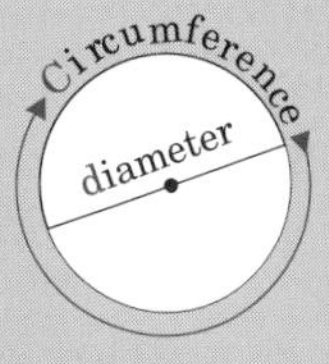

d(3.14) = C

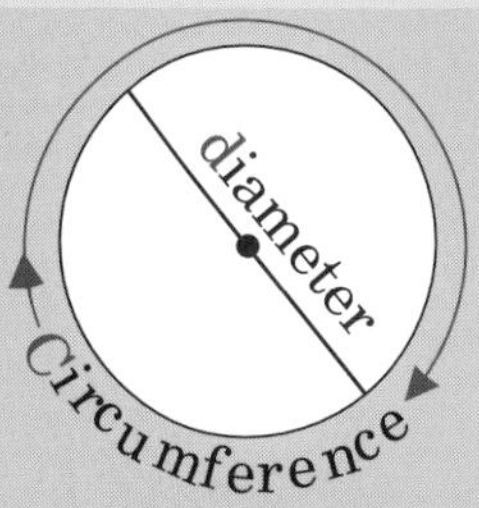

d(3.14) = C

- The number 3.14 has a special symbol: π.
  It's called **pi.**

- Here are two equations for working with circles: **d x = C**
- Both equations say the same thing: **x d = C**
  3.14 times diameter equals circumference. **3.14 x d = C**

Part 4

Figure out the circumference of each circle.

π x d = C

a.

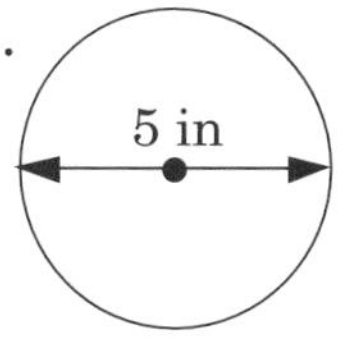

b.

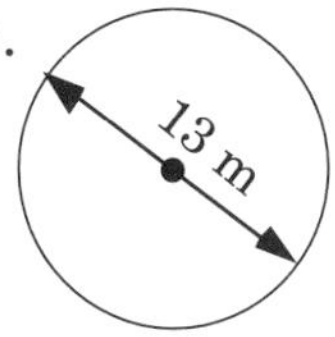

c.

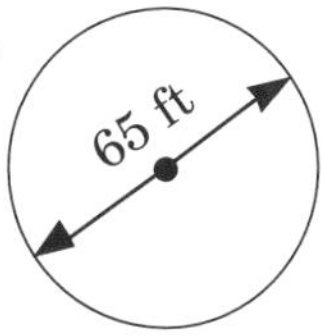

d.

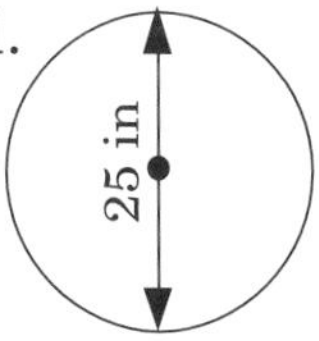

Part 5

Copy each fraction. Write the prime factors for the numerator and the denominator. Cross out any fractions that equal 1. Write the value you get when you multiply.

a. $\frac{40}{50}$ b. $\frac{6}{24}$ c. $\frac{6}{25}$ d. $\frac{1}{14}$

Part 6

- This problem can't be worked the way it is written because the denominators are not the same.
- To work the problem, you find the **lowest common denominator.** That's the first common number your reach when you count by the denominators.

$$\begin{array}{r} \frac{7}{12} \\ + \frac{6}{15} \\ \hline \end{array}$$

- Start with the smaller denominator. That's **12.** Write the numbers for counting by 12.

12, 24, 36, 48, 60, 72 . . .

- Then do the same thing for counting by 15. The first number that shows up for counting by 12 and counting by 15 is the lowest common denominator.

15, 30, 45, 60, 75 . . .

- You can use a calculator to find the first common number.
- Start with [1][2][+][+] . Then each time you press [=], the calculator shows the next number for counting by 12.

	12
[+] [+] [=]	24
[=]	36
[=]	48
[=]	60
[=]	72

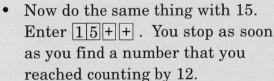

- Now do the same thing with 15. Enter [1][5][+][+] . You stop as soon as you find a number that you reached counting by 12.

	15
[+] [+] [=]	30
[=]	45
[=]	60

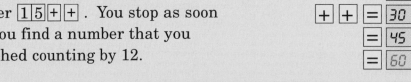

- The first common number is 60. That's the common denominator. You write it for both fractions.

$$\begin{array}{rcr} \frac{7}{12} & = & \frac{\quad}{60} \\ + \frac{6}{15} & = & + \frac{\quad}{60} \\ \hline \end{array}$$

- Then you work the equivalent-fraction problems to figure out the numerators.

$$\begin{array}{rcr} \frac{\mathbf{7}}{\mathbf{12}}\left(\frac{5}{5}\right) & = & \frac{\mathbf{35}}{\mathbf{60}} \\ +\frac{\mathbf{6}}{\mathbf{15}}\left(\frac{4}{4}\right) & = & +\frac{\mathbf{24}}{\mathbf{60}} \\ \hline & & \frac{\mathbf{59}}{\mathbf{60}} \end{array}$$

- Then you add and write the answer.

- Remember the steps:
  - ✔ Start with the smaller denominator.
  - ✔ Write the numbers for counting by that denominator.
  - ✔ Then do the same thing for the larger denominator.
  - ✔ Stop when you find the first number common to both denominators.
  - ✔ Write the common denominator for both fractions.
  - ✔ Work the equivalent-fraction problems.
  - ✔ Then write the answer to the problem.

Part 7

For each problem find the lowest common denominator. Figure out the numerators and work the problem.

a. $$\begin{array}{rcr} \frac{5}{18} & = & \blacksquare \\ -\frac{1}{14} & = & -\blacksquare \\ \hline & & \blacksquare \end{array}$$

b. $$\begin{array}{rcr} \frac{5}{12} & = & \blacksquare \\ +\frac{6}{15} & = & +\blacksquare \\ \hline & & \blacksquare \end{array}$$

Part 8

Copy and work each item.

a. $\frac{7}{10}\left(\frac{\frac{5}{3}}{\frac{5}{3}}\right) = \frac{\blacksquare}{\blacksquare}$

b. $\frac{2}{5}\left(\frac{\frac{1}{9}}{\frac{1}{9}}\right) = \frac{\blacksquare}{\blacksquare}$

Part 9 **For each item, make a number family and a ratio table. Answer the questions.**

a. The tank holds $\frac{7}{5}$ as much as the barrel holds. The barrel holds 65 gallons.

   1. How much does the tank hold?
   2. How much more does the tank hold than the barrel?

b. Michael earned $\frac{2}{3}$ the amount that Walter earned. Walter earned \$145 more than Michael earned.

   1. How much did Walter earn?
   2. How much did Michael earn?

Independent Work

Part 10 Make a fraction number family for each sentence.

a. $\frac{8}{15}$ of the children were girls.

b. 4 out of every 5 cards were black.

c. 3 of every 7 animals were cows.

d. There were $\frac{8}{5}$ as many cards as baskets.

e. House J costs $\frac{9}{2}$ as much as house B.

f. $\frac{3}{11}$ of the children were sick.

Part 11 Copy and complete each equation.

$\angle a = 30°$

$\angle b = \blacksquare$

$\angle k = \blacksquare$

$\angle l = \blacksquare$

$\angle m = \blacksquare$

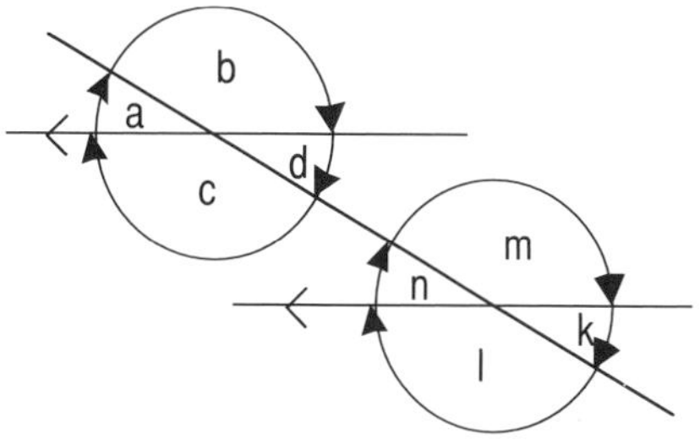

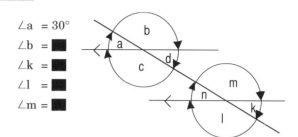

Part 12 For each item, write a ratio equation. Show the answer as a number and a unit name.

a. In Hank Pond, there were 6 water lilies for every 5 frogs. There were 245 frogs in the pond. How many water lilies were there?

b. A machine produced 7 inches of yarn every 2 seconds. How many seconds would it take for the machine to produce 287 inches of yarn?

c. A car traveled at the rate of 90 yards every 8 seconds. The car traveled for 104 seconds. How far did it travel?

d. The ratio of boys to girls at Norman School is 3 to 4. There are 200 girls in Norman School. How many boys are in the school?

Part 13 For each item, make a table and complete it.

a. Multiply each number by 3.
Then add 11.
0, 11, 10, 1

b. Divide each number by 6.
Subtract the answer from 40.
48, 120, 12

Part 14 For each item, figure out what the letter equals.

a. $2M = 56$
$M = \blacksquare = \blacksquare$

b. $3 + 36 - 4 = J$
$J = \blacksquare$

c. $4 \times 42 = P$
$P = \blacksquare$

Part 15 Copy and complete each equation.

a. $\blacksquare = \frac{280}{7}$ b. $\frac{3}{5} = \frac{24}{\blacksquare}$ c. $140 = \frac{\blacksquare}{3}$ d. $11 = \frac{\blacksquare}{36}$ e. $\frac{8}{9} = \frac{\blacksquare}{90}$

Part 16 For each item, write the complete equation. Show the missing value as a fraction.

a. $2(\blacksquare) = 7$ b. $17(\blacksquare) = 190$ c. $4(\blacksquare) = 3$ d. $18(\blacksquare) = 19$

Lesson 56

Part 1 **Copy and complete each equation.**

a. $\frac{3}{7}\left(\frac{\frac{4}{5}}{\frac{4}{5}}\right) = \frac{\blacksquare}{\blacksquare}$

b. $\frac{9}{4}\left(\frac{\frac{1}{6}}{\frac{1}{6}}\right) = \frac{\blacksquare}{\blacksquare}$

c. $\frac{1}{8}\left(\frac{\frac{10}{7}}{\frac{10}{7}}\right) = \frac{\blacksquare}{\blacksquare}$

d. $\frac{5}{3}\left(\frac{\frac{2}{5}}{\frac{2}{5}}\right) = \frac{\blacksquare}{\blacksquare}$

Part 2 **For each item, make a number family and a ratio table. Answer the questions.**

a. $\frac{3}{8}$ of the employees wear glasses. The rest do not. There is a total of 440 employees.
   1. How many employees wear glasses?
   2. How many of the employees do not wear glasses?

b. The regular price for a pair of glasses is $\frac{7}{4}$ the sale price of the glasses. If you buy the glasses on sale, you save $36.
   1. How much is the sale price?
   2. How much is the regular price?

c. The swimming pool is $\frac{9}{10}$ as wide as the garden. The garden is 40 feet wide.
   1. How wide is the swimming pool?
   2. How much wider is the garden than the swimming pool?

d. $\frac{2}{9}$ of the vehicles are trucks. 49 vehicles are not trucks.
   1. What's the total number of vehicles?
   2. How many trucks are there?

Part 3 For each item, figure out the circumference of the circle.

a.

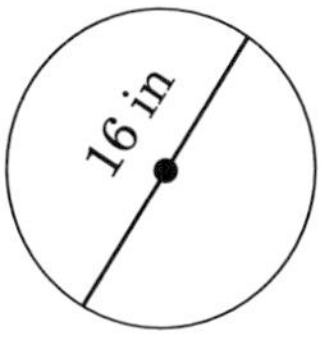

b.

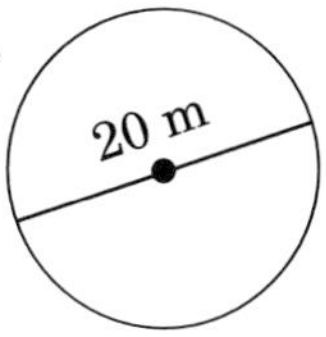

c.

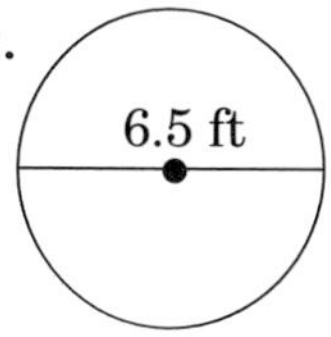

Part 4 For each problem, find the lowest common denominator and figure out the answer.

a. $$\frac{7}{12} - \frac{4}{9} = \blacksquare$$

b. $$\frac{1}{20} + \frac{7}{8} = \blacksquare$$

Part 5 Round each decimal number to the nearest hundredth.

a. 5.1765 b. 5.397 c. .2741 d. 10.495 e. 5.1505 f. 2.4923

Independent Work

Part 6 Simplify each fraction. Show the prime factors.

a. $\frac{36}{48}$ b. $\frac{50}{65}$

c. $\frac{40}{100}$ d. $\frac{24}{48}$

Part 7 Copy each problem. Write the missing value as a fraction.

a. $2(\blacksquare) = 3$ b. $12(\blacksquare) = 1$

c. $1(\blacksquare) = 15$ d. $40(\blacksquare) = 96$

Part 8 Copy and complete the table.

	Multiplication	Fraction equation	Division
a.	$17(\blacksquare) = 51$	$\blacksquare = \blacksquare$	$\overline{)\quad}$
b.			$8\overline{)1128}$

Part 9 Find the area and perimeter of the parallelogram.

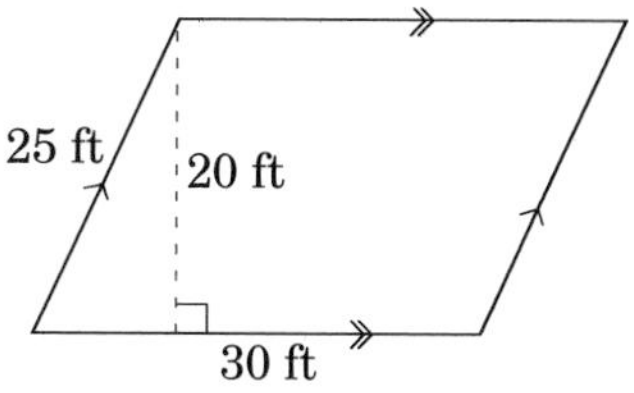

Part 10 **For each item, write a ratio equation. Show the answer as a number and a unit name.**

a. On a car lot, the ratio of cars to trucks is 9 to 2. There are 24 trucks on the lot. How many cars are on the lot?

b. A bug eats 8 leaves every 3 days. How long would it take for the bug to eat 184 leaves?

c. Every 4 bricks weigh 9 pounds. A pile of bricks weighs 1242 pounds. How many bricks are in the pile?

d. The ratio of pigs to goats on a farm is 3 to 2. There are 90 pigs on the farm. How many goats are on the farm?

Part 11 **For each row, write the fraction and the decimal value it equals.**

	Fraction	Decimal
a.	$\frac{456}{10}$	
b.	$\frac{6}{100}$	
c.	$\frac{3}{10}$	
d.	$\frac{148}{100}$	
e.	$\frac{148}{10}$	

Part 12 **Make a number family for each item and answer the question.**

a. A bull weighs 820 pounds more than a cow. The bull weighs 2480 pounds. How many pounds does the cow weigh?

b. $\frac{2}{9}$ of the cows were in the barn. What fraction of the cows were not in the barn?

c. 457 girls attend Lexington School. 395 boys attend Lexington School. How many fewer boys than girls attend Lexington School?

d. 457 girls attend Lexington School. 395 boys attend Lexington School. How many children attend Lexington School?

e. 457 girls attend Lexington School. 395 boys attend Lexington School. What fraction of the children in Lexington School are girls?

Part 13 **Copy and complete each equation.**

a. $25J = 50$ $J = \frac{■}{■} = ■$

b. $M = \frac{207}{3}$ $M = ■$

c. $\frac{4}{9} = \frac{■}{18}$

d. $\frac{6}{7} = \frac{246}{■}$

e. $7 = \frac{■}{26}$

f. $15 = \frac{■}{3}$

Part J

a. $\frac{3}{7}\left(\frac{\frac{4}{5}}{\frac{4}{5}}\right) = \frac{\frac{12}{5}}{\frac{28}{5}}$ b. $\frac{9}{4}\left(\frac{\frac{1}{6}}{\frac{1}{6}}\right) = \frac{\frac{9}{6}}{\frac{4}{6}}$ c. $\frac{1}{8}\left(\frac{\frac{10}{7}}{\frac{10}{7}}\right) = \frac{\frac{10}{7}}{\frac{80}{7}}$ d. $\frac{5}{3}\left(\frac{\frac{2}{5}}{\frac{2}{5}}\right) = \frac{\frac{10}{5}}{\frac{6}{5}}$

Lesson 57

Part 1

For each item, make a number family and a ratio table. Answer the questions.

a. The sale price of the jacket is $\frac{8}{10}$ as much as the regular price. If you buy the jacket on sale, you save $24.

1. How much is the sale price?
2. How much is the regular price?

b. $\frac{5}{9}$ of the students are wearing blue. 28 students are not wearing blue.

1. How many students are wearing blue?
2. How many students are there in all?

c. $\frac{3}{7}$ of the owls haven't produced eggs this year. The rest have. There is a total of 105 owls.

1. How many owls have produced eggs?
2. How many owls haven't produced eggs?

Part 2

If problems have the first number missing, you can figure out that number by working backwards and **undoing** the operation shown in the problem.

- Here are the rules for undoing:
  - ✔ To undo **addition,** you **subtract** the same number.
  - ✔ The undo **subtraction,** you **add** the same number.
  - ✔ To undo **multiplication,** you **divide** by the same number.
  - ✔ To undo **division,** you **multiply** by the same number.

Part 3

For each item, figure out the answer and write the complete equation.

Sample 1	[redacted] + 11 = 34	a.	[redacted] x 12 = 96	b.	[redacted] + 36 = 96
Sample 2	[redacted] x 4 = 40	c.	[redacted] x 46 = 138	c.	[redacted] – 138 = 200

Part 4

Write each item as a column problem and find the common denominator. Then work the equivalent fraction problems to find the answer.

a. $\frac{7}{16} - \frac{3}{40} =$ [redacted]

b. $\frac{3}{14} + \frac{7}{8} =$ [redacted]

Part 5

- You're going to work problems that refer to percent. Percents are related to hundredths decimal numbers. To change a hundredth decimal number into an equivalent percent, remove the decimal point and write the percent sign.

	Decimal	%	
• Here's 56-hundredths:	.56	56%	That's 56 percent.
	1.78	178%	
	5.03	503%	
	.04	4%	

- To go from percents to hundredths, you make sure you have two digits after the decimal point.

5% = .05

75% = .75

671% = 6.71

Part 6

Write the decimal value that equals each percent.

a. 37% b. 8% c. 156% d. 209% e. 99%

Part 7

Write the percent that equals each decimal value.

a. 1.56 b. .75 c. .01 d. .40 e. 3.06

Part 8

For each item, figure out the circumference of the circle.

a.

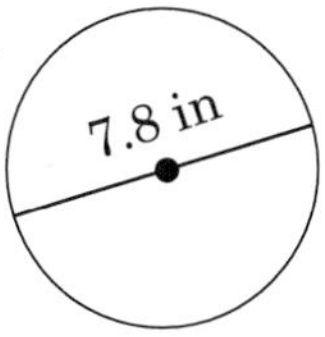

b.

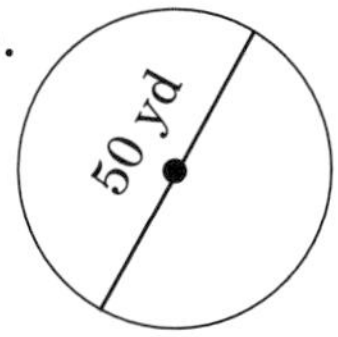

c.

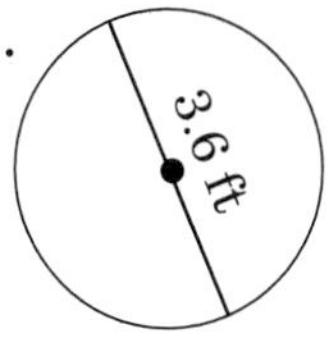

Part 9

Write the fraction that equals 1. Then multiply to complete the equivalent fraction.

a. $\frac{8}{4}\left(\square\right) = \frac{\square}{3}$

b. $\frac{9}{12}\left(\square\right) = \frac{3}{\square}$

c. $\frac{2}{6}\left(\square\right) = \frac{\square}{9}$

d. $\frac{3}{12}\left(\square\right) = \frac{1}{\square}$

Independent Work

Part 10 **For each item, write a ratio equation. Show the answer as a number and a unit name.**

a. A car travels 86 feet every 3 seconds. How many seconds will it take for the car to travel 1032 feet?

b. Every 4 bottles cost $12. How much do 20 bottles cost?

Part 11 **Copy and complete the table.**

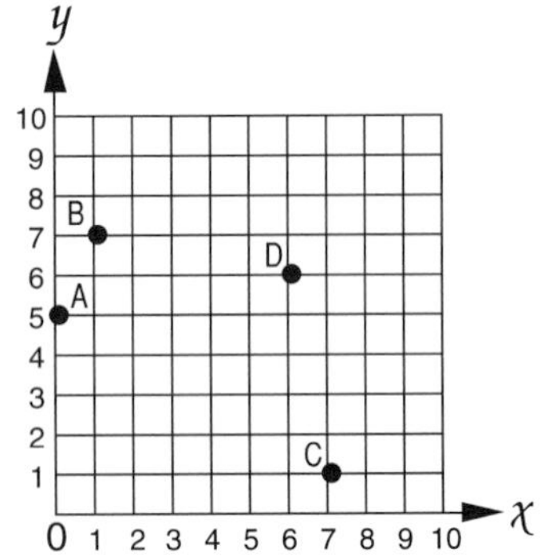

	x	y
A	■	■
B	■	■
C	■	■
D	■	■

Part 12 **Simplify each fraction. Show the prime factors.**

a. $\frac{4}{6}$ b. $\frac{36}{40}$

c. $\frac{18}{24}$ d. $\frac{30}{60}$

Part 13 **Copy each item. Write the fraction each letter equals.**

a. 14F = 6
F = ■

b. 3R = 10
R = ■

c. 56P = 57
P = ■

d. $\frac{2}{8} \times \frac{3}{6} \times \frac{1}{2} = Q$
Q = ■

e. $\frac{1}{4} + \frac{10}{4} + \frac{7}{4} = N$
N = ■

Part 14 **Write the decimal number for each value.**

a. 12 and 7 thousandths

b. 4 and 3 tenths

c. 6 and 45 hundredths

d. 145 and 6 hundredths

Part 15 **Copy and work each problem.**

a. $\frac{5}{9} = \frac{15}{■}$ b. $\frac{■}{9} = 6$

c. $\frac{3}{8} = \frac{■}{48}$ d. $3 = \frac{■}{4}$

Part 16 **For fractions that are more than 1, write an equation to show the fraction and the mixed number or whole number it equals.**

a. $\frac{5}{31}$ b. $\frac{77}{6}$ c. $\frac{17}{20}$ d. $\frac{580}{10}$ e. $\frac{11}{15}$ f. $\frac{18}{17}$ g. $\frac{47}{3}$

Lesson 58

Part 1 **Write the fraction that equals 1. Then multiply to complete the equivalent fraction.**

a. $\frac{3}{10}\left(■\right) = \frac{■}{7}$

b. $\frac{11}{5}\left(■\right) = \frac{2}{■}$

c. $\frac{4}{3}\left(■\right) = \frac{9}{■}$

d. $\frac{8}{9}\left(■\right) = \frac{■}{1}$

Part 2 **For each item, figure out the answer and write the complete equation.**

a. ■ – 48 = 13

b. ■ x 56 = 224

c. ■ ÷ 3 = 24

d. ■ + 80 = 86

e. ■ x 15 = 75

Part 3 **Work each item. If necessary, round your answer to hundredths.**

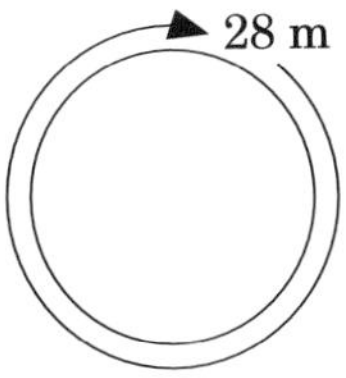

a. Find the diameter.

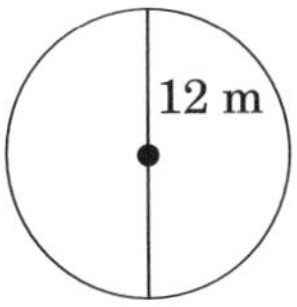

b. Find the circumference.

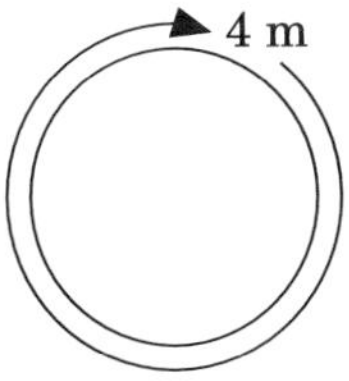

c. Find the diameter.

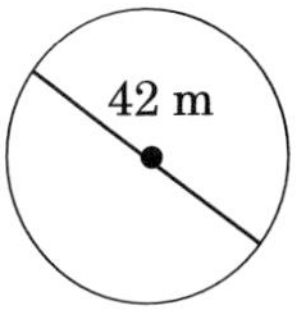

d. Find the circumference.

Part 4 **Write each item as a column problem and find the common denominator. Then work the equivalent fraction problems to find the answer.**

a. $\frac{4}{10} - \frac{4}{15} =$ ■

b. $\frac{7}{8} + \frac{1}{10} =$ ■

Part 5 **Copy and complete the table.**

	Fraction	Decimal	%
a.	$\frac{19}{100}$		
b.	$\frac{385}{100}$		
c.	$\frac{100}{100}$		
d.	$\frac{8}{100}$		
e.	$\frac{805}{100}$		

Part 6 For each item, make two equations. Answer the question.

> ***Sample problem***
>
> You start with 4 and multiply by 3. Then you add 5. What number do you end up with?
>
> 4 x 3 = ■ ?
>
> ■ + 5 = ■

a. You start with 26 and add 19. Then you divide by 5. What number do you end up with?

b. You start with 76 and subtract 20. Then you multiply by 4. What number do you end up with?

c. You start with 344 and divide by 4. Then you multiply by 3. What number do you end up with?

Independent Work

Part 7 Copy and complete the table.

	Fraction	Decimal	Mixed number
a.	$\frac{136}{100}$		
b.		3.08	
c.			$4\frac{3}{10}$
d.		17.015	

Part 8 For each item, write a ratio equation and answer the question.

a. Joe ran at the rate of 16 yards every 5 seconds. How many yards did Joe run in 30 seconds?

b. A tractor uses 5 gallons of fuel to plow every 4 acres. How many gallons does the tractor need to plow 72 acres?

c. On a ranch, the sheep ate 16 bushels of feed for every 9 bushels of feed the cows ate. If the cows ate 81 bushels of feed, how much did the sheep eat?

Part 9 Copy each table. Figure out the missing numbers.

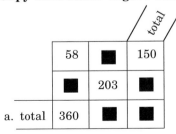

			total
	58	■	150
	■	203	■
a. total	360	■	■

			total
	23	■	■
	■	32	■
b. total	66	■	99

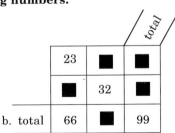

Part 10

For each item, make a number family with a difference number. Answer each question.

a. Mittens cost $2.68 less than the gloves cost. How much do mittens cost?

b. Sue needs $9.00 to buy the boots. How much does Sue have?

Part 11

For each fraction that is more than 1, write the complete equation to show the fraction and the whole number or mixed number it equals.

a. $\frac{116}{5}$ b. $\frac{75}{10}$ c. $\frac{40}{45}$ d. $\frac{48}{2}$ e. $\frac{50}{205}$

Part 12

For each item, make a number family and answer each question.

a. Hill Road is 19 miles shorter than Beach Road. Hill Road is 74 miles long. How long is Beach Road?

b. Fred weighs 124 pounds. His horse weighs 1340 pounds. How much do Fred and his horse weigh together?

Part 13

Copy each problem and write the answer.

a. $\frac{4}{3} - \frac{3}{3} = \blacksquare$

b. $\frac{12}{20} + \frac{5}{20} = \blacksquare$

c. $\blacksquare = \frac{200}{20}$

d. $\frac{\blacksquare}{7} = 7$

e. $11(\blacksquare) = 6$

f. $\frac{20}{11} = \frac{\blacksquare}{33}$

g. $\frac{40}{41} = \frac{120}{\blacksquare}$

Part 14

Copy and work each problem.

a. $19\overline{)94}$ b. $82\overline{)501}$

Lesson 59

Part 1

For each item, figure out the answer and write the complete equation.

a. ■ ÷ 6 = 68

b. ■ – 203 = 140

c. ■ x 56 = 672

d. ■ + 291 = 314

Part 2

For each item, write the fraction that equals 1. Then multiply to complete the equivalent fraction.

a. $\frac{3}{7}\left(\blacksquare\right) = \frac{\blacksquare}{6}$ ■

b. $\frac{8}{4}\left(\blacksquare\right) = \frac{10}{\blacksquare}$ ■

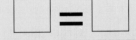

c. $\frac{10}{5}\left(\blacksquare\right) = \frac{\blacksquare}{4}$ ■

Part 3

- An **equal sign** shows that the values on both sides of the sign are **equal.** □ = □
- An **inequality sign** is used to show that the sides are **not equal.** > or <
- The side that is closer to the **bigger end** of the inequality sign is the side with the **bigger value.** □ > □ (bigger value, smaller value)
- The side closer to the **pointed end** of the sign is the side with the **smaller value.**

Part 4

Copy each item. Write the appropriate sign to show which side has the larger value or whether the sides are equal.

a. $1 \blacksquare \frac{7}{6}$ b. $\frac{10}{1} \blacksquare 2$ c. $\frac{10}{5} \blacksquare 2$

d. $\frac{10}{5} \blacksquare \frac{12}{5}$ e. $9 \blacksquare \frac{36}{9}$ f. $\frac{13}{14} \blacksquare \frac{14}{13}$

Part 5

Write each item as a column problem and find the common denominator. Then work the equivalent fraction problems to find the answer.

a. $\frac{5}{12} + \frac{9}{16} = \blacksquare$ b. $\frac{11}{12} - \frac{8}{9} = \blacksquare$

Part 6

For each item, make three equations. Answer the question.

a. You start with 44 and divide by 4. Then you add 209. Then you divide by 10. What number do you end up with?

b. You start with 800 and divide by 4. Then you subtract 200. Then you multiply by 6. What number do you end up with?

c. You start with 65 and multiply by 7. Then you add 50. Then you divide by 5. What number do you end up with?

Part 7

Copy the table. Fill in the missing values.

		$\frac{\blacksquare}{100}$	Decimal	%
a.	$\frac{2}{5}$			
b.	$\frac{5}{4}$			
c.	$\frac{1}{2}$			
d.	$\frac{11}{10}$			

Independent Work

Part 8 **Make a ratio table. Answer the questions in each item. Write your answer as a number and a unit name.**

a. Bernard has 2 U.S. stamps for every 7 foreign stamps. Bernard has 105 foreign stamps.

1. How many U.S. stamps does Bernard have?
2. How many stamps does he have in all?

b. In the school library, there are hardback books and paperback books. The ratio of hardback books to total books is 5 to 9. There are a total of 720 books in the library.

1. How many hardback books are there?
2. How many paperback books are there?

Part 9 **Find the circumference of each circle.**

a.

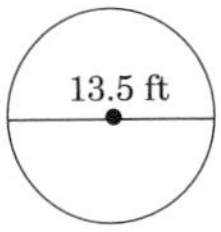

b.

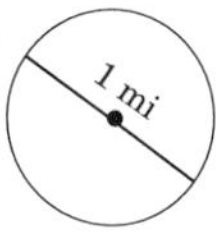

c.

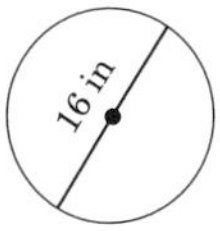

Part 10 **Copy and complete the table to show the x and y values for each point.**

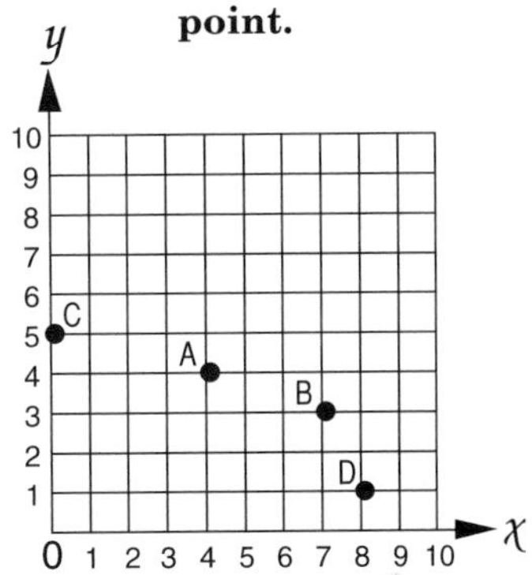

	x	y
A	■	■
B	■	■
C	■	■
D	■	■

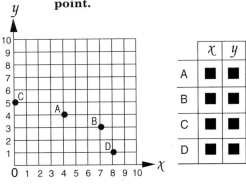

Part 11 **Copy the table and complete it.**

	Division	Multiplication
a.		14 (■) = 70
b.	$8\overline{)296}$	

Part 12 **Make a table. Fill in all the missing numbers. Answer each question.**

Paul and his brother Mike collect National League and American League baseball cards.

Facts

1. Paul has 752 American League cards.
2. Mike has 624 American League cards.
3. Mike has a total of 952 cards.
4. Paul has 269 National League cards.

Items

a. Mike has 212 more stamps than American League baseball cards. How many stamps does he have?

b. Jesse also has a card collection. He has 550 fewer cards than Paul's total collection. How many cards does Jesse have?

Part 13 **Copy and complete each item.**

a. $4(■) = 5$ b. $18(■) = 1$ c. $3F = 5$ d. $2J = 7$

$F = ■ = ■$ $J = ■ = ■$

Part 14 **Make a fraction number family for each item. Box the answer to the question.**

a. $\frac{4}{5}$ of the animals were sheep. What fraction of the animals were not sheep?

b. 4 of the cards were jacks. There were 52 cards. What fraction of the cards were not jacks?

c. $\frac{2}{9}$ of the cars were dirty. What is the fraction for all the cars?

Part 15 **Simplify each fraction.**

a. $\frac{9}{12}$ b. $\frac{15}{20}$ c. $\frac{42}{48}$ d. $\frac{16}{20}$

Part J c. $\frac{10}{5} = 2$ d. $\frac{10}{5} < \frac{12}{5}$ e. $9 > \frac{36}{9}$ f. $\frac{13}{14} < \frac{14}{13}$

Mike, I think there's an easier way to figure out the answer.
stamps cards
624
dif cards stamps
212

Lesson 60

Part 1

- You've worked with fractions that can be simplified.
- Some mixed numbers have fractions that can be simplified.
- Here's: $5\frac{2}{4}$
- To simplify this mixed number, we keep the whole number and simplify the fraction.

$$\frac{2}{4} = \frac{\cancel{2}}{\cancel{2} \times 2} = \frac{1}{2}$$

- So the simplified mixed number for $5\frac{2}{4}$ is $5\frac{1}{2}$.

$$5\frac{2}{4} = 5\frac{1}{2}$$

Part 2

Show each mixed number with a simplified fraction. If the fraction cannot be simplified, copy the mixed number the way it is written.

a. $13\frac{8}{16}$ b. $4\frac{11}{35}$ c. $3\frac{18}{36}$ d. $4\frac{9}{28}$ e. $7\frac{14}{35}$ f. $14\frac{72}{81}$

Part 3

Work each item. If necessary, round your answer to hundredths.

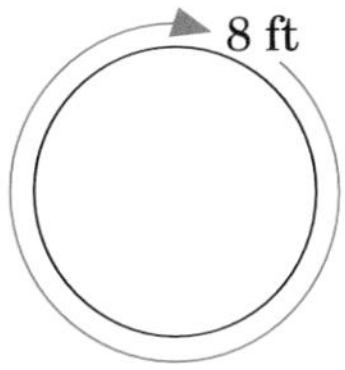

a. Find the diameter.

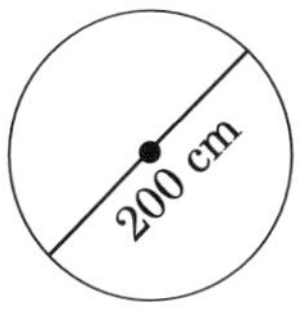

b. Find the circumference.

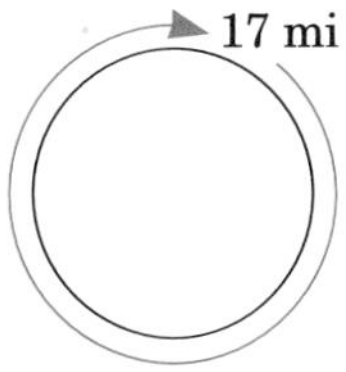

c. Find the diameter.

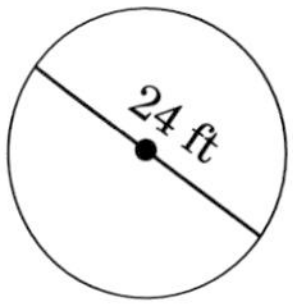

d. Find the circumference.

Part 4 Copy the table. Fill in the missing values.

		$\frac{\blacksquare}{100}$	Decimal	%
a.	$\frac{9}{20}$			
b.	$\frac{7}{5}$			
c.	$\frac{3}{2}$			
d.	$\frac{1}{4}$			

Part J

b. $\pi \times d = C$
$3.14 \times 200 = C$
C = 628 cm

c. $\pi \times d = C$
$3.14 \left(\frac{17}{3.14}\right) = 17$
d = 5.41 mi

d. $\pi \times d = C$
$3.14 \times 24 = C$
C = 75.36 ft

Test 6

Part 1 Show the prime factors for each item. Write the simplified fraction.

a. $\frac{12}{30}$ b. $\frac{7}{28}$ c. $\frac{18}{35}$ d. $\frac{40}{60}$

Part 2 For each problem, make a fraction number family and a ratio table. Then answer the questions.

a. The regular price of a shirt is $\frac{7}{5}$ of the sale price. The regular price is $35.
   1. What is the difference between the sale price and the regular price?
   2. What is the sale price?

b. $\frac{4}{9}$ of the flowers bloomed. There were 108 flowers.
   1. How many flowers did not bloom?
   2. How many flowers bloomed?

Part 3

Write the x and y values for each point.

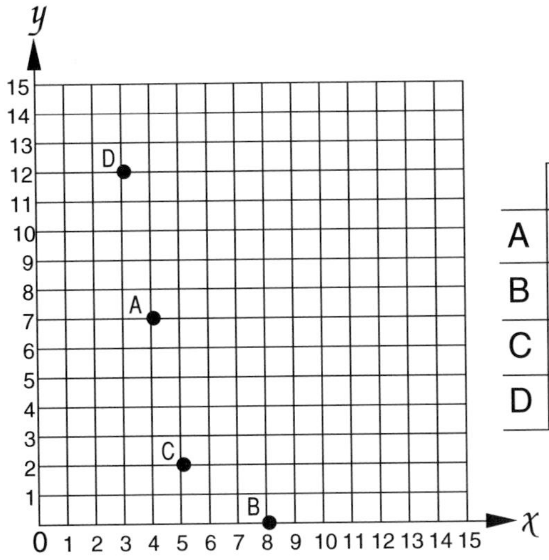

	x	y
A	■	■
B	■	■
C	■	■
D	■	■

Part 4

Write the equation to show the fraction that equals F and the whole number or mixed number it equals.

a. 3F = 74

b. 8F = 13

c. 5F = 69

Part 5

Copy and complete each equation.

a. ■ x 12 = 84

b. ■ – 200 = 56

Part 6

Complete each equation. Show the answer as a fraction.

a. $\frac{3}{8}\left(■\right) = \frac{7}{■}$

b. $\frac{4}{3}\left(■\right) = \frac{■}{14}$

Part 7

Write each problem in a column. Find the common denominator. Box the answer.

a. $\frac{11}{8} + \frac{4}{6} =$ ■

b. $\frac{15}{16} - \frac{3}{10} =$ ■

Part 8

Use your calculator to work each problem. Write the answer as a mixed number.

a. $15 \div 4 =$

b. $27 \div 8 =$

c. $400 \div 32 =$

Lesson 61

Part 1 **Copy each item. Write the appropriate sign: >, < or = .**

a. $\frac{13}{13} \blacksquare \frac{7}{7}$ b. $\frac{5}{8} \blacksquare \frac{8}{5}$ c. $1 \blacksquare \frac{60}{60}$ d. $\frac{65}{64} \blacksquare \frac{64}{65}$

Part 2 **Write each mixed number in its most simplified form.**

a. $3\frac{55}{100}$ b. $2\frac{8}{10}$ c. $7\frac{8}{19}$ d. $4\frac{21}{49}$

Part 3

- You can compare some pairs of fractions by examining them. Other pairs are more difficult.
- You can compare those fractions by finding the lowest common denominator. When the denominators are the same, the fraction with the larger numerator is the larger fraction.
- Here's a problem: **Which is more, $\frac{3}{4}$ or $\frac{5}{8}$?**
- We write the fractions in a column and figure out the lowest common denominator.
- When both fractions have a denominator of 8, you can see that the larger fraction is $\frac{6}{8}$.

$$\frac{3}{4}\left(\frac{2}{2}\right) = \frac{6}{8}$$
$$\frac{5}{8} = \frac{5}{8}$$

$\frac{3}{4} = \frac{6}{8}$ **so:** $\boxed{\frac{3}{4} > \frac{5}{8}}$

Part 4 **Work each item.**

a. Which is more, $\frac{2}{3}$ or $\frac{3}{5}$?

b. Which is more, $\frac{9}{12}$ or $\frac{5}{6}$?

Part 5 **Copy and complete each equation.**

a. $\frac{7}{4}\left(\blacksquare\right) = \frac{9}{\blacksquare}$

b. $\frac{8}{3}\left(\blacksquare\right) = \frac{\blacksquare}{2}$

c. $\frac{5}{9}\left(\blacksquare\right) = \frac{\blacksquare}{5}$

d. $\frac{6}{10}\left(\blacksquare\right) = \frac{1}{\blacksquare}$

Part 6

- This problem shows a mystery number. The question mark above the **first box** shows that the first number is the mystery number.

$$\overset{?}{\blacksquare} \times 2 = \blacksquare$$
$$\blacksquare + 4 = 64$$

- To work problems of this type, you start at the **end** of the second equation and work **backward.** You **undo** each step that was done.

- We work the problem a step at a time.
- First we undo adding 4:

$$\overset{?}{\blacksquare} \times 2 = \boxed{60}$$
$$\boxed{60} + 4 = 64$$

$$64 - 4 = \boxed{60}$$

- Then we undo multiplying by 2:

$$\overset{?}{\blacksquare} \times 2 = \boxed{60}$$
$$\boxed{60} + 4 = 64$$

$$2\overline{)60} \quad ?$$

Part 7 **Figure out the mystery number in each item.**

a. $\overset{?}{\blacksquare} + 12 = \blacksquare$
$\blacksquare \times 3 = 48$

b. $\overset{?}{\blacksquare} + 210 = \blacksquare$
$\blacksquare - 58 = 400$

c. $\overset{?}{\blacksquare} \times 5 = \blacksquare$
$\blacksquare + 16 = 76$

Part 8 Work each item. If necessary, round your answer to hundredths.

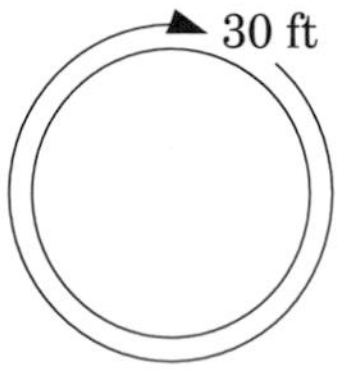

a. Find the diameter.

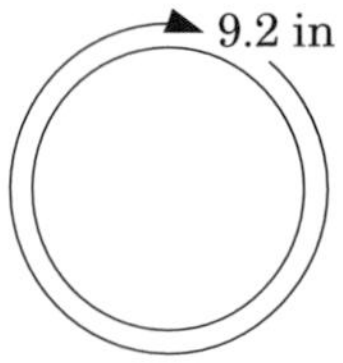

b. Find the diameter.

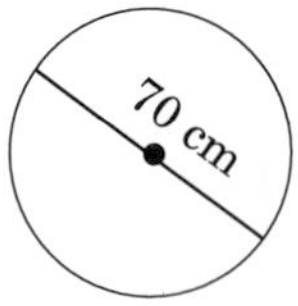

c. Find the circumference.

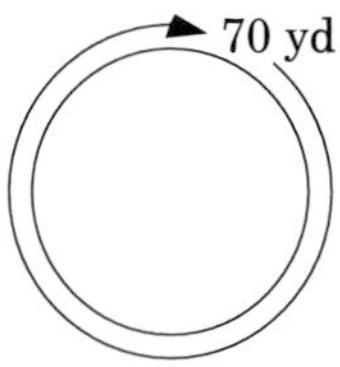

d. Find the diameter.

Part 9 For each fraction, write the equivalent hundredths fraction. Then write the percent it equals.

Sample

$$\frac{3}{5}\left(\frac{20}{20}\right) = \frac{60}{100} = 60\%$$

a. $\frac{10}{5}$ b. $\frac{1}{2}$ c. $\frac{3}{4}$ d. $\frac{12}{10}$ e. $\frac{4}{25}$

Independent Work

Part 10 For each problem, make a fraction number family. Answer the questions.

a. $\frac{3}{7}$ of the children wore white socks. What fraction of the children did not wear white socks?

b. A building had 84 windows. 66 of them were dirty. What fraction of the windows were not dirty?

c. 23 of the cards were blue. The rest were white. There were 50 cards in all.
   1. What's the fraction for the cards that were white?
   2. What's the fraction for the cards that were blue?

Part 11 Work each problem.

a. $\frac{3}{8} + \frac{3}{4} = \blacksquare$

b. $\frac{7}{8} - \frac{1}{6} = \blacksquare$

c. $\frac{3}{15} + \frac{2}{10} = \blacksquare$

Part 12 Copy each equation. Write each answer as a whole number or mixed number.

a. $\frac{246}{9} = \blacksquare$ b. $\frac{128}{8} = \blacksquare$ c. $\frac{584}{73} = \blacksquare$

Part 13 **For each problem, make a fraction number family and a ratio table. Answer the questions.**

a. $\frac{2}{3}$ of fish in a pond were bass. The pond had 150 fish that were not bass.

1. How many bass were in the pond?
2. How many fish were in the pond?

b. Jan ran $\frac{2}{5}$ of the distance to Bean Town. The total distance to Bean Town was 40 miles. How far did Jan run?

Part 14 **For each item, write the equation to show the mixed number and the fraction it equals.**

a. $10\frac{3}{5}$ = [redacted] b. $36\frac{1}{2}$ = [redacted] c. $9\frac{4}{11}$ = [redacted] d. $1\frac{3}{50}$ = [redacted] e. $12\frac{5}{8}$ = [redacted]

Part J

a. $\frac{13}{13} = \frac{7}{7}$ b. $\frac{5}{8} < \frac{8}{5}$ c. $1 = \frac{60}{60}$ d. $\frac{65}{64} > \frac{64}{65}$

Part K

b. $\pi \times d = C$
$3.14\left(\frac{9.2}{3.14}\right) = 9.2$
d = 2.93 in

c. $\pi \times d = C$
$3.14 \times 70 = C$
C = 219.8 cm

d. $\pi \times d = C$
$3.14\left(\frac{70}{3.14}\right) = 70$
d = 22.29 yd

BEAN TOWN
40 miles
Cliff, I think there's a faster way to figure out how far Jan ran.

Lesson 62

Part 1

- Some ratio problems require a table. Those are problems that have **three names.** Problems that have only two names do not require a table.
- Here's a problem that requires a table:

> The ratio of girls to children is 1 to 3. If there are 60 children, how many boys are there?

girls	■	■
boys	■	■
children	■	■

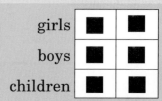

- The names are **girls, boys** and **children.** The problem requires a table because it has three names.
- Here's a problem that does not require a table:

> The ratio of girls to children is 2 to 5. If there are 20 girls, how many children are there?

$$\frac{\text{girls}}{\text{children}}\ \frac{■}{■}\left(\frac{■}{■}\right)=\frac{■}{■}$$

- The names are **girls** and **children.**
- Remember, if there are **three names,** you need a table. If there are only two names, you don't need a table. Also remember that one of the names may be in the question.

Part 2

Read each problem. Figure out whether or not you need to make a table. Then make the ratio equation or the table.

a. There are 3 sunny days for every 4 days that are not sunny. If there are 84 total days, how many are sunny days?

b. The machine makes 11 garments every 3 hours. How many hours will it take for the machine to make 99 garments?

c. In a pond, 3 of every 4 fish are bass. If there are 90 fish that are not bass, how many fish are in the pond?

Part 3

Answer each question.

a. Which is larger, $\frac{6}{5}$ or $\frac{8}{7}$?

b. Which is larger, $\frac{3}{4}$ or $\frac{5}{6}$?

Part 4

Rewrite each boxed answer as a mixed number or a whole number.

a. $= \frac{18}{\boxed{\frac{340}{9}}}$

b. $= \frac{35}{\boxed{\frac{91}{7}}}$

c. $= \frac{\boxed{\frac{99}{5}}}{9}$

d. $= \frac{\boxed{\frac{295}{8}}}{24}$

Part 5

- Some ratio problems are difficult because you have to figure out a complicated fraction that equals 1.
- Here's a problem:

  There are 5 pies for every 8 muffins. If there are 11 muffins, how many pies are there?

- The first sentence tells the names and the numbers:

$$\frac{\text{pies}}{\text{muffins}}\ \frac{5}{8}$$

- The rest of the problem tells us that we have 11 muffins. We have to figure out how many pies we have. We write the parentheses and equal sign and 11 for muffins.

$$\frac{\text{pies}}{\text{muffins}}\ \frac{5}{8}\left(\quad\right) = \frac{}{\boxed{11}}$$

- We have two numbers on the bottom, so we can work that problem: 8 x (▮) = 11. That's $\frac{11}{8}$.

$$\frac{\text{pies}}{\text{muffins}}\ \frac{5}{8}\left(\frac{\frac{11}{8}}{\frac{11}{8}}\right) = \frac{}{11}$$

- So the fraction that equals 1 is: $\frac{\frac{11}{8}}{\frac{11}{8}}$.

- Now we work the problem on top. The answer is $\frac{55}{8}$ pies. That's almost 7 pies.

$$\frac{\text{pies}}{\text{muffins}}\ \frac{5}{8}\left(\frac{\frac{11}{8}}{\frac{11}{8}}\right) = \frac{\boxed{\frac{55}{8}}}{11}$$

Part 6

Make an equation with names and numbers. Then figure out the answer to the question.

a. The ratio of perch to trout is 5 to 7. If there are 9 perch, how many trout are there?

b. There are 5 potatoes needed for every 2 dinners. If there are 7 dinners, how many potatoes are needed?

c. There were 8 servings for every 6 pounds of meat. How many pounds of meat were used for 10 servings?

d. The ratio of planks to cabinets is 4 to 3. How many planks are there if there are 7 cabinets?

Part 7

Copy each problem. Figure out the mystery number.

a.
$$\overset{?}{■} - 100 = ■$$
$$■ \times 4 = 340$$

b.
$$\overset{?}{■} - 97 = ■$$
$$■ - 11 = 31$$

Part 8

- You've worked circumference problems that tell about the diameter or ask about the diameter.
- Some problems tell about the **radius** or ask about the **radius.**

The radius is $\frac{1}{2}$ the diameter.

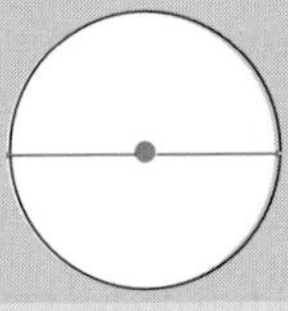

- To calculate the **radius,** you first find the **diameter.** Then you divide by 2.
- If the problem **tells about the radius,** you multiply the radius by 2 to get the diameter. Then you work the problem.
- Remember, the radius is $\frac{1}{2}$ the diameter.

Part 9 Work each item. If necessary, round your answer to hundredths.

a. 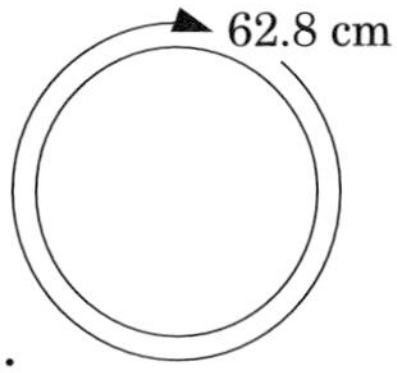

r = ■

b. 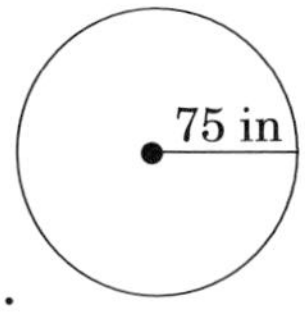

C = ■

c. 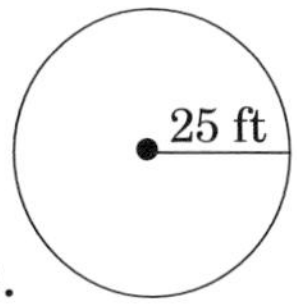

C = ■

d. 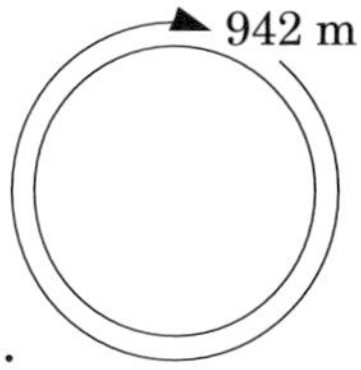

r = ■

Independent Work

Finish the problems in part 2.

Part 10 Make a fraction number family for each item. Answer the questions.

a. 38 people were on the bus. 20 of them were children.
   1. What fraction of the people were not children?
   2. What fraction of the people were children?

b. 5 of the goats had long horns. The rest did not. There were 19 goats in all.
   1. What fraction did not have long horns?
   2. What fraction did have long horns?

c. $\frac{3}{23}$ of the desks had 9 drawers. What fraction of the desks did not have 9 drawers?

Part 11 Find the area and the perimeter of each figure.

a. 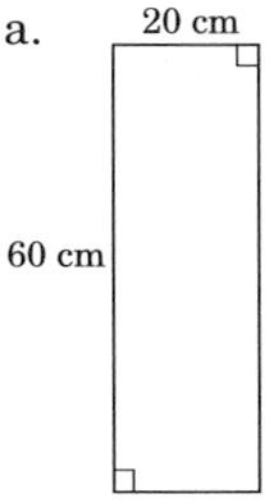

b. 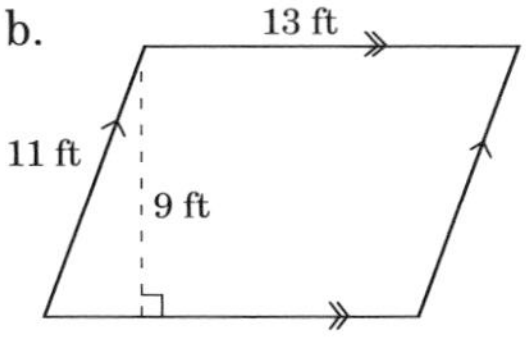

c. 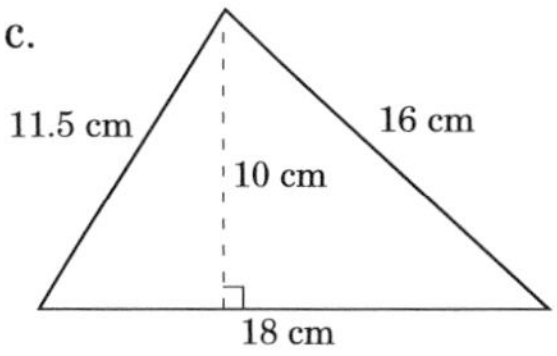

Part 12 Work each problem.

a. $\frac{3}{5} + \frac{1}{5} + \frac{7}{5} = \blacksquare$

b. $\frac{3}{5} \times \frac{1}{5} \times \frac{3}{2} = \blacksquare$

c. $7 = \frac{\blacksquare}{5}$

d. $\frac{\blacksquare}{10} = 12$

e. $\frac{8}{7} = \frac{40}{\blacksquare}$

f. $\frac{3}{9} = \frac{\blacksquare}{36}$

g. $\frac{4}{5}\left(\blacksquare\right) = \frac{16}{25}$

h. $\frac{3}{7}\left(\blacksquare\right) = \frac{21}{70}$

i. $\frac{3}{8}\left(\frac{4}{5}\right) = \blacksquare$

Part 13 Copy and complete each equation.

$\angle k = 108°$

$\angle b = \blacksquare$

$\angle d = \blacksquare$

$\angle c = \blacksquare$

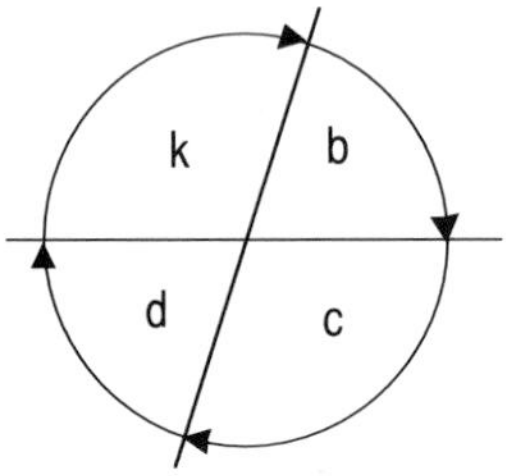

Lesson 63

Part 1 **For each item, write a complicated fraction that equals 1. Write the answer to the question as a fraction. Then figure out the whole number or mixed number the fraction equals.**

a. A mixture uses 6 parts of silver for every 11 parts of gold. There are 52 parts of silver in the mixture. How many parts of gold are there?

b. There were 5 birds for every 2 ounces of birdseed. If there was a total of 88 birds, how much birdseed was there?

Part 2 **Answer each question.**

a. Which is more, $\frac{3}{8}$ or .4?

b. Which is more, .12 or $\frac{1}{6}$?

c. Which is more, .6 or $\frac{4}{7}$?

Part 3 **Copy each problem. Figure out the mystery number.**

a. (? above the first box)
■ − 100 = ■
■ x 3 = 75

b. (? above the first box)
■ x 2 = ■
■ − 420 = 10

Part 4

- You can figure out the answer to all ratio problems by showing the fraction that equals 1 as a fraction over a fraction. But for some problems, it's faster if you don't.
- You look at the problem you start with.
- If you know that the answer to that problem is a whole number, you just write a simple fraction.

- Here's a problem: $\frac{5}{9}\left(\quad\right) = \frac{35}{\square}$
- The fraction that equals 1 is a simple fraction. $\frac{5}{9}\left(\frac{7}{7}\right) = \frac{35}{\square}$

- Here's another problem: $\frac{5}{9}\left(\frac{\quad}{\quad}\right) = \frac{425}{\square}$
- You know the missing value is a whole number, but you don't know what number it is. So you could write this fraction: $\frac{425}{5}$. $\frac{5}{9}\left(\frac{\frac{425}{5}}{\quad}\right) = \frac{425}{\square}$
- Or you could figure out what 425 ÷ 5 equals. $\frac{5}{9}\left(\frac{85}{\quad}\right) = \frac{425}{\square}$

- Here's another problem: $\frac{8}{5}\left(\quad\right) = \frac{\square}{3}$
- You know the missing value is not a whole number. So you write this fraction: $\frac{3}{5}$. $\frac{8}{5}\left(\frac{\quad}{\frac{3}{5}}\right) = \frac{\square}{3}$
- The fraction that equals one is: $\frac{\frac{3}{5}}{\frac{3}{5}}$. $\frac{8}{5}\left(\frac{\frac{3}{5}}{\frac{3}{5}}\right) = \frac{\square}{3}$

Part 5 Copy and work each problem.

a. $\frac{10}{4}(\blacksquare) = \frac{\blacksquare}{94}$

b. $\frac{3}{7}(\blacksquare) = \frac{18}{\blacksquare}$

c. $\frac{7}{3}(\blacksquare) = \frac{91}{\blacksquare}$

d. $\frac{9}{8}(\blacksquare) = \frac{45}{\blacksquare}$

e. $\frac{9}{6}(\blacksquare) = \frac{\blacksquare}{92}$

f. $\frac{12}{5}(\blacksquare) = \frac{\blacksquare}{50}$

Part 6 For each item, write table or no table. Then make the ratio equation or the table.

a. In Grosvenor City, the summer has sunny days and days that are not sunny. There are 3 sunny days for every 5 days that are not sunny. If there are 66 sunny days, how many days are not sunny?

b. The ratio of girls to boys is 6 to 5. If there are 30 boys, how many girls are there?

c. At a school, there are students and employees. The ratio of employees to total people is 4 to 9. If there are 603 total people in the school, how many are employees? How many are students?

Part 7 Work each item. If necessary, round your answer to hundredths.

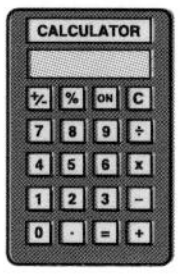

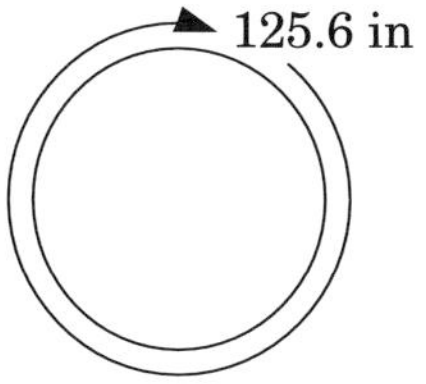

a. Find the radius.

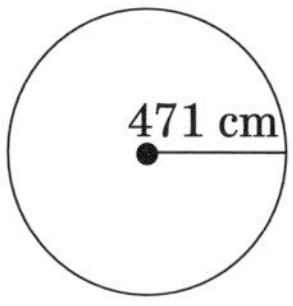

b. Find the circumference.

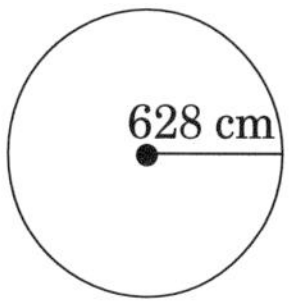

c. Find the circumference.

Independent Work

Finish the problems in part 6.

Part 8 **Make a fraction number family for each item. Answer the questions.**

a. The road went 17 miles. 14 miles were through hills. What fraction of the road was not through hills?

b. $\frac{3}{7}$ of the doors in a building squeaked. What fraction of the doors did not squeak?

c. 14 sheets of paper were white. 17 sheets were colored.

1. What's the fraction for the sheets that were colored?
2. What's the fraction for all the sheets?

Part 9 **For each item, make a fraction number family and a ratio table. Answer the questions.**

a. The cost of the chair was $\frac{8}{5}$ the cost of the stool. The stool cost $35.

1. What was the cost of the chair?
2. How much more was the cost of the chair than the cost of the stool?

b. $\frac{2}{3}$ of the people were sleeping. There were 90 people in all.

1. How many were sleeping?
2. How many were awake?

Part 10 **Copy and complete each equation. If the answer is 1 or more, show it as a whole number or mixed number.**

a. 5R = 6

R = ■ = ■

b. 12R = 1

R = ■

c. 25J = 11

J = ■

d. 7R = 7

R = ■ = ■

e. 2P = 64

P = ■ = ■

f. 9J = 8

J = ■

Part 11 **Copy the table and write the x and y values for each point.**

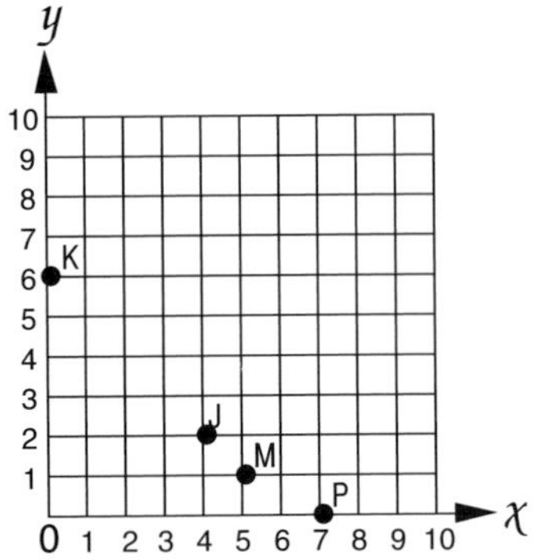

	x	y
J	■	■
K	■	■
M	■	■
P	■	■

Part 12 **Copy and complete the table.**

	Fraction	Decimal	Mixed number
a.	$\frac{364}{100}$		
b.		12.02	
c.			$17\frac{2}{10}$
d.	$\frac{23}{10}$		

Part 13 **Find the area and perimeter of each figure.**

a.

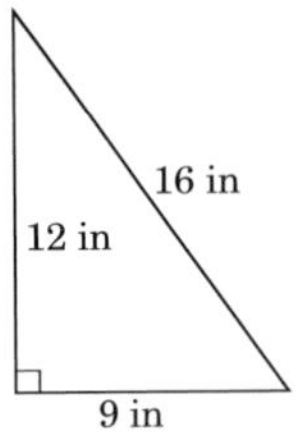

b.

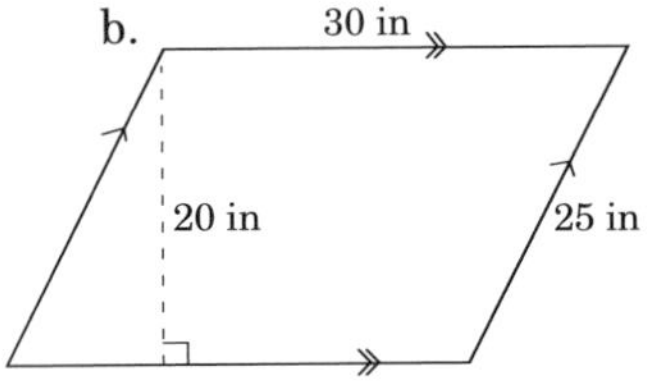

Part 14 **Copy and work each problem.**

a. $\frac{13}{2}\left(■\right) = \frac{26}{26}$ b. $\frac{3}{5} = \frac{15}{■}$ c. $\frac{2}{3} = \frac{■}{18}$

d. $3 = \frac{■}{14}$ e. $■ = \frac{29}{4}$ f. $18 = \frac{■}{4}$

Part J

b. $\frac{3}{7}\left(\frac{6}{6}\right) = \frac{18}{\boxed{42}}$

c. $\frac{7}{3}\left(\frac{\frac{91}{7}}{\frac{91}{7}}\right) = \frac{91}{\boxed{\frac{273}{7}}}$ $\boxed{39}$ $\left[\text{or } \frac{7}{3}\left(\frac{13}{13}\right) = \frac{91}{\boxed{39}}\right]$

d. $\frac{9}{8}\left(\frac{5}{5}\right) = \frac{45}{\boxed{40}}$

e. $\frac{9}{6}\left(\frac{\frac{92}{6}}{\frac{92}{6}}\right) = \frac{\boxed{\frac{828}{6}}}{92}$ $\boxed{138}$

f. $\frac{12}{5}\left(\frac{10}{10}\right) = \frac{\boxed{120}}{50}$

If there are 7 doors in the building and none of them squeak, what fraction of the doors squeak?

Lesson 64

Part 1 **For each item, write the decimal value as a fraction. Then find the lowest common denominator. Write the statement that compares the original values.**

a. Which is more, $\frac{2}{8}$ or .25?

b. Which is more, $\frac{5}{9}$ or .5?

c. Which is more, $\frac{3}{4}$ or .74?

Part 2 **For each item, figure out the mystery number.**

a. ■ (?) x 2 = ■
■ – 64 = 100

b. ■ (?) – 501 = ■
■ x 3 = 12

c. ■ (?) + 28 = ■
■ – 100 = 50

Part 3 **Work each item. If necessary, round your answer to hundredths.**

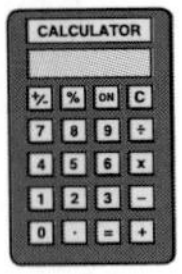

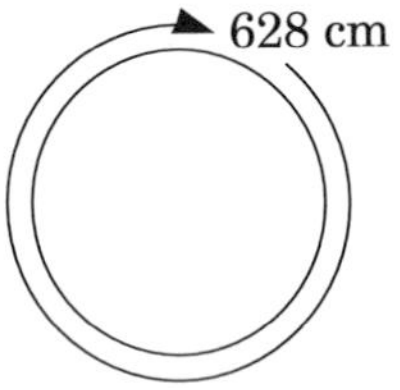

a. Find the diameter.

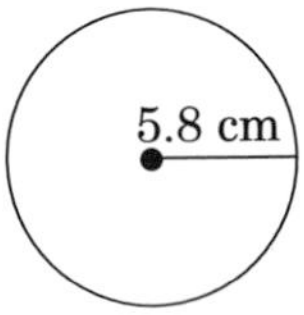

b. Find the circumference.

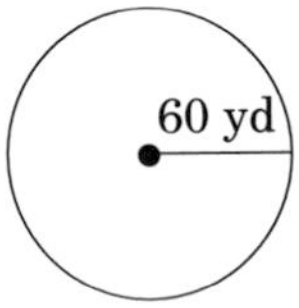

c. Find the circumference.

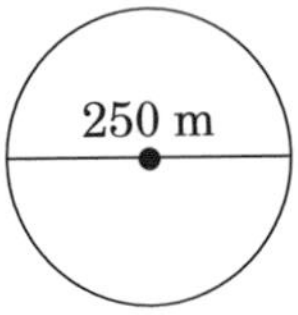

d. Find the circumference.

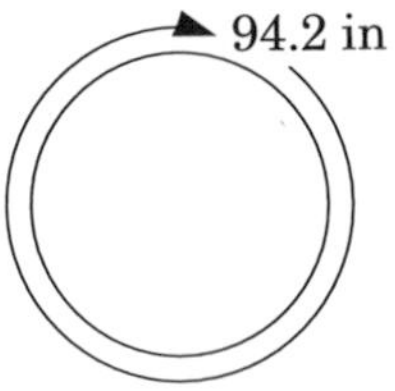

e. Find the radius.

Part 4 **For each item, write two equations. Figure out the mystery number.**

a. You start with the mystery number and subtract 18. Then you multiply by 5. You end up with 45. What's the starting number?

b. You start with the mystery number and multiply by 9. Then you add 15. You end up with 69. What's the starting number?

c. You start with the mystery number and subtract 20. Then you multiply by 4. You end up with 84. What's the starting number?

d. You start with the mystery number and multiply by 3. Then you multiply by 5. You end up with 105. What's the starting number?

Part 5 **Work each ratio problem. Then write the answer as a mixed number and a unit name.**

a. If 6 bottles fill 1 container, how many containers do 43 bottles fill?

b. Jimmy walked 24 miles in 7 hours. He walked at a steady pace. How far did Jimmy walk in 6 hours?

c. 5 containers are the same size. Together, the 5 containers hold 18 ounces. How many ounces do 8 of those containers hold?

d. If it takes 8 hours to paint 6 rooms, how many rooms can be painted in 14 hours?

Part 6 For each problem, make the ratio equation or the table.

a. In a factory, there are 4 blue buttons for every 5 red buttons. If there are 2000 red buttons, how many blue buttons are there?

b. In a company, there are 3 secretaries for every 4 departments. There are 48 departments. How many secretaries are there?

c. A factory uses red buttons and yellow buttons. The ratio of red buttons to yellow buttons is 3 to 7. If the factory uses 3000 total buttons, how many red buttons are used? How many yellow buttons are used?

d. In a store, 3 out of every 11 employees are cashiers. A store has 121 employees. How many employees are not cashiers? How many employees are cashiers?

Independent Work

Finish the problems in part 6.

Part 7 Make a fraction number family for each item. Answer the questions.

a. A building had 77 doors. 12 of them needed repair. What fraction of the doors did not need repair?

b. In a classroom, there were 15 girls and 12 boys.
   1. What's the fraction for all the students?
   2. What's the fraction for the boys?

c. There were 21 coins. 5 of them were new.
   1. What fraction was not new?
   2. What fraction was new?

Part 8 Copy and work each problem by finding the lowest common denominator.

a. $\frac{7}{8} + \frac{1}{2}$

b. $\frac{13}{20} - \frac{1}{6}$

Part 9 For each item, write the decimal value as a fraction and find the lowest common denominator. Then write the statement that compares the original values.

a. Which is more, .24 or $\frac{2}{8}$?

b. Which is more, $\frac{7}{12}$ or .6?

Part 10 Copy and work each problem.

a. $\frac{13}{5} - \frac{13}{5} =$ ■ b. $\frac{3}{8} \times \frac{1}{2} \times \frac{3}{2} =$ ■ c. $\frac{2}{17} \times \frac{17}{2} =$ ■ d. $\frac{1}{5} + \frac{1}{5} + \frac{5}{5} =$ ■

e. $\frac{1}{7} = \frac{3}{■}$ f. $\frac{3}{5}$(■) $= \frac{33}{50}$ g. $\frac{5}{7} = \frac{■}{56}$

Part 11 Copy the table and write the x and y values for each point.

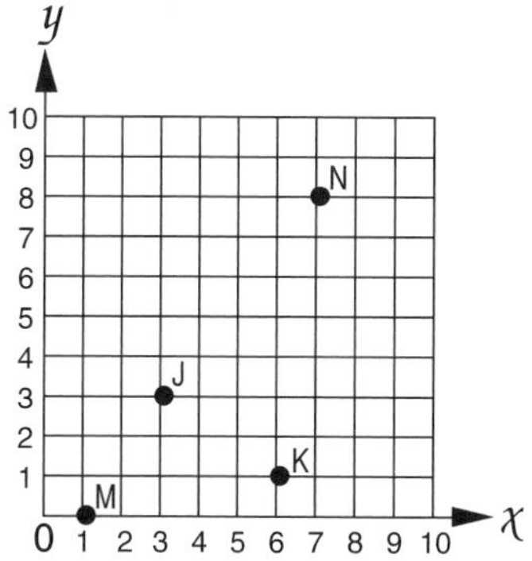

	x	y
J	■	■
K	■	■
M	■	■
N	■	■

Part 12 Copy and complete the table.

		$\frac{■}{100}$	Decimal	%
a.	$\frac{4}{5}$			
b.	$\frac{8}{5}$			
c.	$\frac{22}{4}$			
d.	$\frac{2}{25}$			

Part 13 Copy and complete each equation.

∠k = 19°
∠t = ■
∠p = ■
∠r = ■

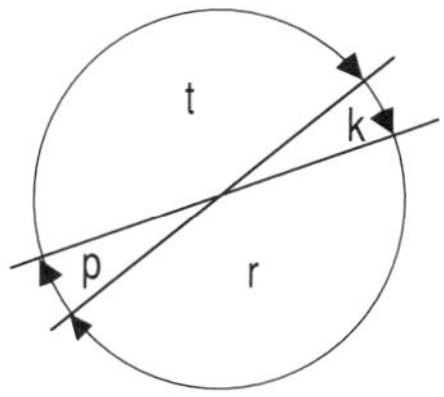

Part J

a. $\frac{2}{8}\left(\frac{25}{25}\right) = \frac{50}{200}$ $\frac{25}{100}\left(\frac{2}{2}\right) = \frac{50}{200}$ $\frac{2}{8} = .25$

b. $\frac{5}{9}\left(\frac{10}{10}\right) = \frac{50}{90}$ $\frac{5}{10}\left(\frac{9}{9}\right) = \frac{45}{90}$ $\frac{5}{9} > .5$

c. $\frac{3}{4}\left(\frac{25}{25}\right) = \frac{75}{100}$ $\frac{74}{100} = \frac{74}{100}$ $\frac{3}{4} > .74$

Part K

a. $\frac{\text{blue}}{\text{red}}$ $\frac{4}{5}$ ( ) = $\frac{□}{2000}$

b. $\frac{\text{secretaries}}{\text{departments}}$ $\frac{3}{4}$ ( ) = $\frac{□}{48}$

c.

red	3	
yellow	7	
buttons		3000

d.

cashiers	3	
not cashiers		
employees	11	121

Lesson 65

Part 1

For each item, write the values in a column. Find the common denominator. Write the answer as a mixed number.

a. $\frac{1}{2} + 3 + \frac{3}{4} =$ [redacted]

b. $2 + \frac{3}{10} + \frac{1}{5} =$ [redacted]

c. $\frac{1}{4} + \frac{1}{6} + 1 =$ [redacted]

Part 2

- When you multiply by a value, you may end up with more than you started with. You may end up with the same value you started with, or you may end up with less than you started with.
- Here are the rules:
  - ✔ If you end up with more than you started with, the value you multiply by is more than 1.
  - ✔ If you end up with the same value you started with, the value you multiply by is 1.
  - ✔ If you end up with less than you started with, the value you multiply by is less than 1.

Part 3

For each problem, write whether the missing value is more than 1 or less than 1 using the symbols > or <. Then write the missing value as a fraction.

Sample
90 x M = 91

a. 70 x M = 17

b. 2 x M = 1

c. 5 x M = 123

d. 56 x M = 11

Part 4

For each problem, write which value is larger, R or M.

a. R x 1.03 = M

b. R x $\frac{30}{29}$ = M

c. R x .2 = M

d. R x $\frac{3}{1}$ = M

e. R x $\frac{1}{3}$ = M

f. R x $\frac{17}{12}$ = M

Part 5

Find the missing side for each rectangle. Use the equation b x h = A.

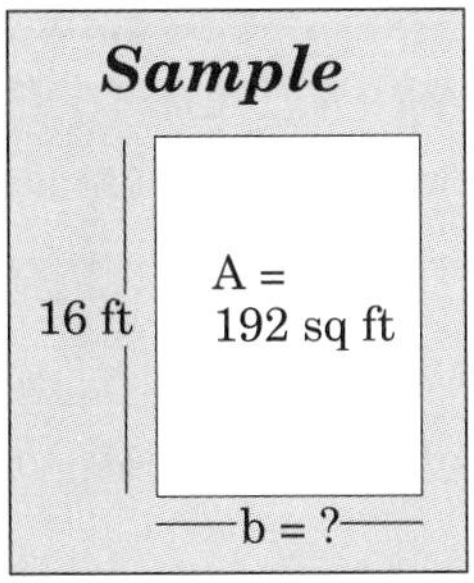

a.

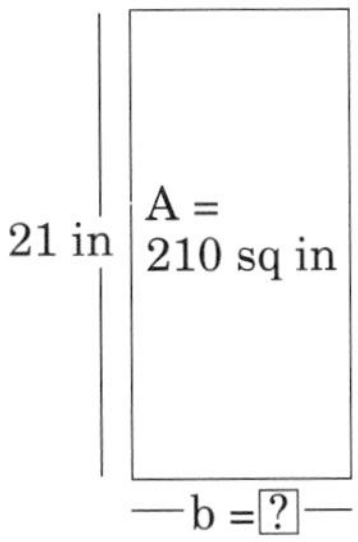

b.

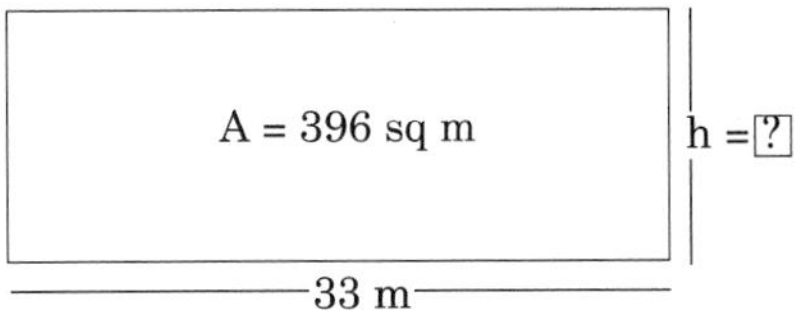

Part 6

Work each ratio problem. Then write the answer as a mixed number or a whole number and a unit name.

a. 8 bags hold 5 pounds. How many bags contain 41 pounds?

b. A factory makes 7 tractors in 2 days. How long will it take to make 21 tractors?

c. A train travels at a steady rate. If the train goes 128 miles in 3 hours, how far does the train go in 2 hours?

Part 7

For each item, write two equations. Figure out the mystery number.

a. You start with a mystery number. You add 52. Then you multiply by 2. You end up with 678. What's the starting number?

b. You start with a mystery number. You divide by 7. Then you add 77. You end up with 84. What's the starting number?

c. You start with a mystery number and multiply by 16. Then you subtract 506 and end up with 86. What's the starting number?

d. You start with a mystery number and you multiply by 30. Then you add 280. You end up with 310. What's the starting number?

Independent Work

Part 8 **Write a ratio table or a ratio equation for each problem. You may have to make a fraction number family for some of the ratio-table problems.**

a. The cost of the book was $\frac{3}{7}$ the cost of the jacket. The book cost $20 less than the jacket.
   1. What was the price of the jacket?
   2. What was the price of the book?

b. 2 out of 3 babies in a hospital were crying. There were 28 babies who were crying. How many babies were in the hospital?

c. For every 5 men, 2 men wore red hats. 30 men wore red hats. How many men did not wear red hats?

d. In a second-hand store, the cost of the TV was $\frac{8}{3}$ the cost of the chair. The cost of the TV was $60 more than the cost of the chair.
   1. How much did the chair cost?
   2. How much did the TV cost?

Part 9 Answer the questions.

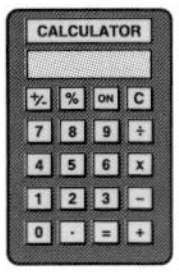

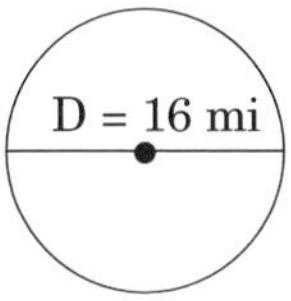

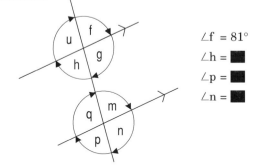

a. What's the circumference?

b. What's the radius?

Part 10 Copy and complete the table.

	$\frac{■}{100}$	Decimal	%
a. $\frac{3}{50}$			
b. $\frac{2}{10}$			
c. $\frac{7}{4}$			
d. $\frac{5}{2}$			

Part 11 Simplify each fraction.

a. $\frac{3}{6} =$ b. $\frac{12}{14} =$ c. $\frac{16}{20} =$

Part 12 Figure out angle t.

∠r = 36°

∠b = 74°

∠t = ■

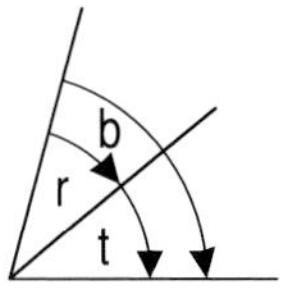

Part 13 For each item, write both values as fractions and find the lowest common denominator. Then write the statement that compares the original values.

a. Which is more, $\frac{4}{15}$ or $\frac{7}{25}$?

b. Which is more, $\frac{11}{30}$ or .37?

Part 14 Copy and complete each equation.

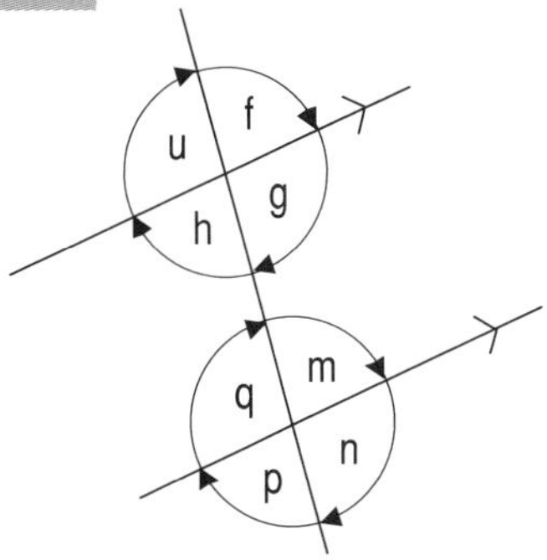

∠f = 81°

∠h = ■

∠p = ■

∠n = ■

Part 15 Copy and complete each table.

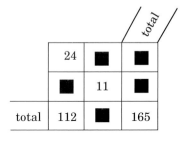

			total
	24	■	■
	■	11	■
total	112	■	165

Part J

c. 37 × 16 = 592 ?
592 − 506 = 86

d. 1 × 30 = 30 ?
30 + 280 = 310

Lesson 66

Part 1

For each item, write whether M is more or less than 1 using the symbols > or <. Then write what M equals.

a. $35 \times M = 11$　　b. $5 \times M = 6$　　c. $24 \times M = 23$

Part 2

For each item, write the values in a column. Find the common denominator. Write the answer as a mixed number.

a. $\frac{1}{4} + \frac{1}{5} + 1 = \blacksquare$　　b. $\frac{3}{8} + \frac{3}{2} + 3 = \blacksquare$

Part 3

For each problem, write N or T to show which is larger.

a. $N \times 9.9 = T$

b. $N \times .99 = T$

c. $N \times \frac{7}{8} = T$

d. $N \times \frac{1}{2} = T$

e. $N \times \frac{3}{2} = T$

f. $N \times \frac{5}{4} = T$

g. $N \times \frac{4}{5} = T$

Part 4

- Some word problems use the word **each.** The word **each** means **1.** 45 miles **each** hour means 45 miles in **1** hour. If they make 156 cars **each** day, they make 156 cars in **1** day.
- You have worked problems that tell about **each** or **1** as multiplication problems or division problems. You can also work them as ratio problems.
- Here's a problem:

There were 8 ounces in each can. If there were 648 ounces, how many cans were there?

$$\frac{\text{oz}}{\text{cans}} \quad \frac{8}{1}\left(\quad\right) = \frac{648}{\blacksquare}$$

Part 5

Work each ratio problem. Answer the question the problem asks.

a. Each tub holds 280 gallons. How many gallons do 6 tubs hold?

b. There were 6 spots on each ladybug. If there were 384 spots, how many ladybugs were there?

c. There were 144 cookies in 6 boxes. How many cookies were in each box?

d. Each bottle holds 5 pints of oil. If there are 72 pints of oil, how many bottles are filled?

Part 6

For each item, write two equations. Figure out the mystery number.

a. You start with a number and multiply by 7. Then you divide by 5. You end up with 14. What's the mystery number?

b. You start with a number. You divide by 3. Then you divide by 8. You end up with 72. What's the mystery number?

c. You start with a number. You add 100. Then you divide by 5. You end up with 35. What number did you start with?

Part 7

Figure out the missing side for each rectangle.

a.

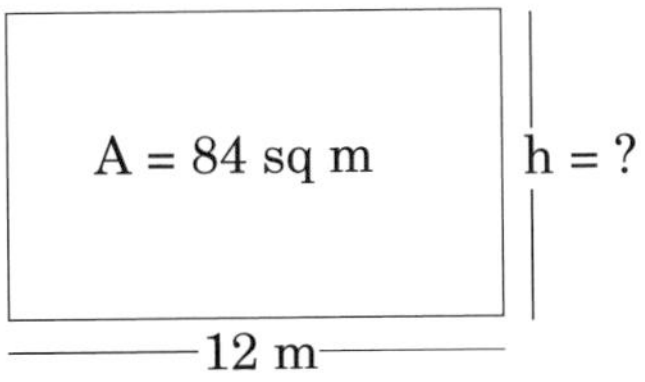

b.

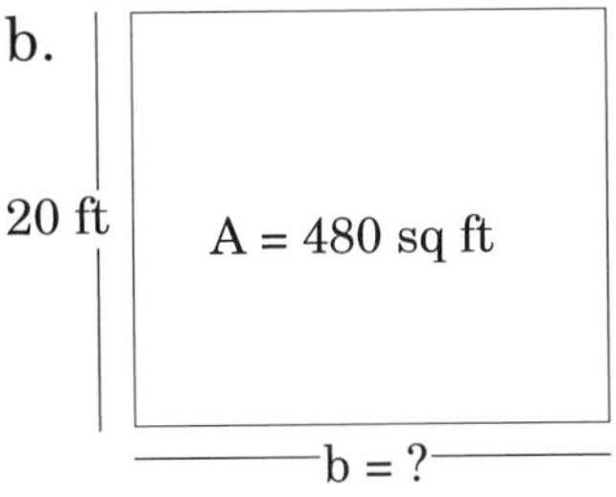

c.

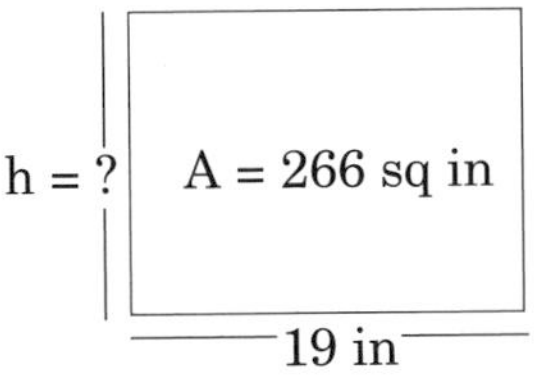

Independent Work

Part 8 **Work each problem. Make a ratio table if you need one.**

a. The number of workers that wore glasses was $\frac{9}{4}$ the number that did not wear glasses. 72 workers wore glasses.

1. How many workers did not wear glasses?
2. How many more workers wore glasses than did not wear glasses?

b. The ratio of sand to cement was 7 to 2. The total mixture weighed 342 pounds.

1. How many pounds of sand were in the mixture?
2. How many pounds of cement were in the mixture?

c. The ratio of sand to cement was 5 to 2. If 600 pounds of sand were used, how many pounds of cement were used?

d. $\frac{3}{8}$ of the students in the school caught chicken pox. 125 students did not catch chicken pox.

1. How many students caught chicken pox?
2. How many students were there in all?

e. $\frac{5}{16}$ of the workers wore steel-toed shoes. 99 workers did not wear steel-toed shoes.

1. How many total workers were there?
2. How many workers wore steel-toed shoes?

Part 9 **Answer each question.**

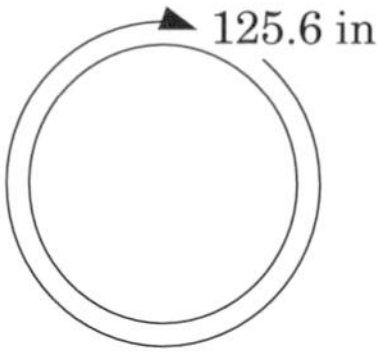

a. What's the radius?

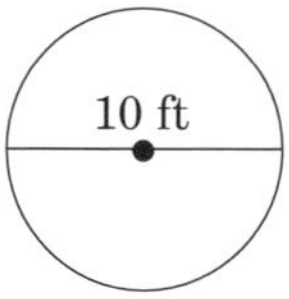

b. What's the circumference?

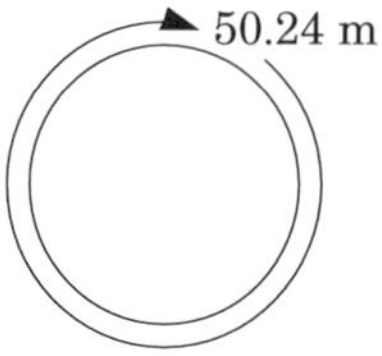

c. What's the radius?

Part 10 **Copy the table and write the x and y values for each point.**

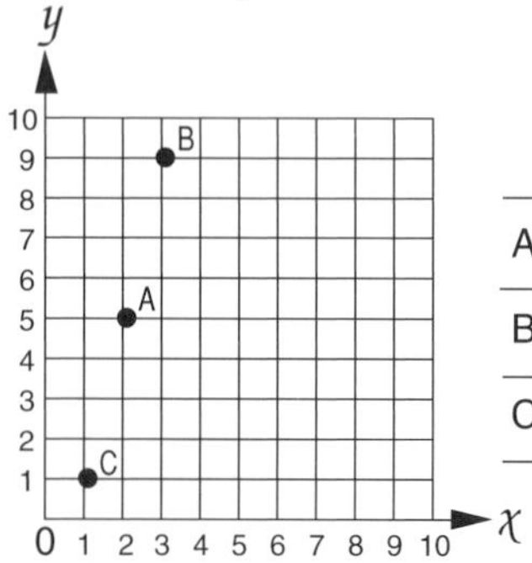

	x	y
A	■	■
B	■	■
C	■	■

Part 11 **Make a number family for each problem. Answer the questions.**

a. The Brinks building had 874 dirty windows and 521 clean windows. How many fewer clean windows than dirty windows were there?

b. The Brinks building had 874 dirty windows and 521 clean windows. How many windows where there in all?

c. The Brinks building had 874 dirty windows and 521 clean windows.

1. What fraction of the windows were dirty?
2. What fraction of the windows were clean?

Lesson 67

Part 1 **For each item, write the equations. Figure out the mystery number.**

a. You start with some number. Then you multiply by 5. Then you subtract 16. Then you add 11. Then you divide by 3. You end up with 30. What's the mystery number?

b. You start with a number. Then you divide by 3. Then you subtract 20. Then you multiply by 2. You end up with 8. What's the mystery number?

c. You start with a number. Then you add 56. Then you divide by 7. Then you multiply by 20. You end up with 160. What's the mystery number?

Part 2

- You've worked with word problems that use the word **each** to mean **1. Each** box is **1** box.
- Sometimes the word **per** is used just like **each.**
- Here's a problem:

  A machine makes 36 baseball cards in 3 seconds. How many **cards per second** does the machine make?

  - Here's the ratio: $\frac{\text{cards}}{\text{sec}} \quad \frac{36}{3}\left(\quad\right) = \frac{\blacksquare}{1}$
- The last fraction has **1** for seconds.

Part 3 **Work each ratio problem. Answer the question the problem asks.**

a. A train travels at a steady 64 miles per hour. How far does the train travel in 8 hours?

b. There are 120 students in 5 classes. How many students per class are there?

c. A farmer drops 4 seeds into each hole. If he plants 136 seeds, how many holes does he need?

Part 4

Write the correct answer for each item. Then check the answer on your calculator.

a. 43 (■) = 38.27

.89 or 1.12

b. 56.1 (■) = 50.49

.9 or 1.1

c. 90 (■) = 91.8

.98 or 1.02

d. 52 (■) = 114.4

.22 or 2.2

Part 5

Work each item. Show the answer as a whole number or mixed number.

a. $3 \times \frac{4}{6} \times \frac{7}{8} =$ ■

b. $3 + \frac{1}{6} + \frac{7}{8} =$ ■

c. $\frac{15}{2} - 3 =$ ■

d. $\frac{2}{5} \times 10 \times \frac{1}{2} =$ ■

Part 6

Make a table and fill in all the missing numbers. Then answer each question.

The only vehicles on Jones Road and Creek Road were cars and trucks.

Facts

1. The total number of trucks for both roads was 87.
2. There were 64 cars on Jones Road.
3. On Creek Road, there were 94 cars.
4. On Jones Road, there were 35 fewer trucks than cars.

Items

a. On which road were there more trucks?

b. What was the total number of cars?

c. What was the total number of vehicles on Creek Road?

Independent Work

Part 7

For each item, indicate whether R or M is larger.

a. R x 2.3 = M b. R x .23 = M c. R x .903 = M d. R x .03 = M e. R x 1.2 = M

Part 8

For some problems, write a fraction number family. For some problems, write a ratio table or a ratio equation. Answer the questions.

a. $\frac{5}{8}$ of the students were boys. What fraction of the students were girls?

b. There were 66 dogs. 17 of them had short tails. What fraction of the dogs had long tails?

c. 12 babies were sleeping. 41 were awake.
   1. What's the fraction for all the babies?
   2. What's the fraction for the sleeping babies?

d. A pie started out with 8 slices. 2 slices of the pie were removed.
   1. What's the fraction for the slices that were not removed?
   2. What's the fraction for the slices that were removed?

e. The fir tree was $\frac{8}{5}$ the height of the poplar. The difference in the height of the trees was 33 feet.
   1. How tall was the fir?
   2. How tall was the poplar?

f. The train moved at $\frac{5}{3}$ the speed of the bus. The bus averaged 45 miles per hour.
   1. What was the speed of the train?
   2. How much faster was the train than the bus?

g. There were 6 short-tailed dogs for every 15 long-tailed dogs. There were 105 long-tailed dogs. How many short-tailed dogs were there?

h. A machine made 3 shirts every 4 minutes. How many minutes would it take the machine to make 7 shirts?

i. A recipe used 4 potatoes for every 3 onions. How many onions would be needed for 5 potatoes?

j. $\frac{2}{5}$ of the doctors wore glasses. If there were 65 doctors, how many wore glasses?

Part 9

For each problem, write the equations for the missing angles.

a.

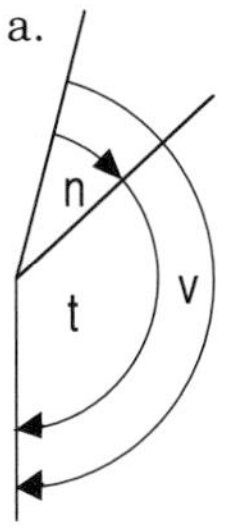

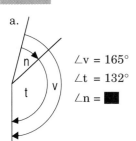

$\angle v = 165°$
$\angle t = 132°$
$\angle n =$ ■

b.

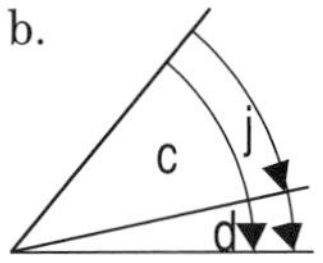

$\angle j = 50°$
$\angle d = 12°$
$\angle c =$ ■

c.

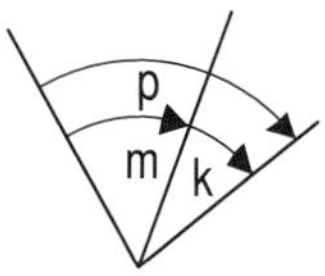

$\angle m = 49°$
$\angle k = 12°$
$\angle p =$ ■

d.

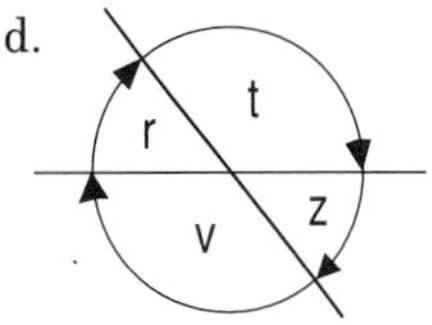

$\angle r = 51°$
$\angle t =$ ■
$\angle v =$ ■
$\angle z =$ ■

Part J

a. $\frac{3}{1} \times \frac{4}{6} \times \frac{7}{8} = \frac{84}{48} = 1\frac{36}{48}$

b.
$$3 = \frac{72}{24}$$
$$\frac{1}{6}\left(\frac{4}{4}\right) = \frac{4}{24}$$
$$+\ \frac{7}{8}\left(\frac{3}{3}\right) = +\frac{21}{24}$$
$$\frac{97}{24} = 4\frac{1}{24}$$

c.
$$\frac{15}{2} = \frac{15}{2}$$
$$-3 = -\frac{6}{2}$$
$$\frac{9}{2} = 4\frac{1}{2}$$

d. $\frac{2}{5} \times \frac{10}{1} \times \frac{1}{2} = \frac{20}{10} = 2$

Lesson 68

Part 1 Figure out the missing value for each rectangle.

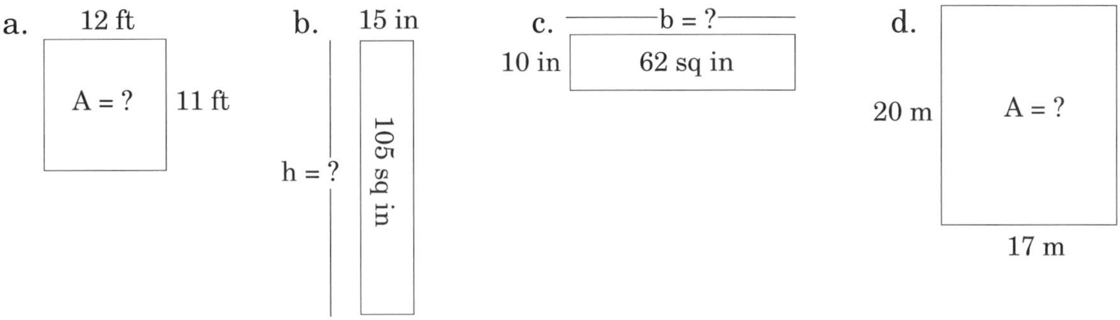

Part 2 For each item, figure out which choice is correct. Check your choice using your calculator.

a. .56 x $\blacksquare$ = .728

1.3 or .77

b. 3.2 x $\blacksquare$ = 2.88

1.11 or .9

c. 7.5 x $\blacksquare$ = 7.65

1.02 or .98

d. 6.1 x $\blacksquare$ = 6.222

1.02 or .98

e. 4.9 x $\blacksquare$ = 4.802

1.02 or .98

f. 3.5 x $\blacksquare$ = 3.43

1.02 or .98

Part 3 **Work each problem as a ratio problem. Remember to show the answer as a number and a unit name.**

a. There are 8 tablets per box. How many boxes are needed for 512 tablets?

b. Each tub holds 61 gallons of water. How many tubs are needed for 549 gallons?

c. Each engine has 6 cylinders. If there are 234 cylinders, how many engines are there?

d. If it takes a machine 5 seconds to make 6 buttons, how many seconds does it take the machine to make each button?

e. A train travels 580 miles in 7 hours. How many miles does the train travel each hour?

Part 4

- You've worked ratio problems that involve fractions. You make a fraction number family.
- You can do the same thing for problems that use decimal values or percents. You make a fraction number family.
- Here is a sentence that gives a decimal value:

  .3 of the children had new shoes.

- Here's the fraction number family with .3 written as a fraction:

new	not new	shoes
$\frac{3}{10}$	$\frac{7}{10}$ →	$\frac{10}{10}$

- Here's a sentence that gives a percent value:

  The rainfall in August was 40% of the rainfall in July.

- Here's the fraction number family with 40% written as a fraction:

dif	August	July
$\frac{60}{100}$	$\frac{40}{100}$ →	$\frac{100}{100}$

Part 5 **For each problem, make a fraction number family and a ratio table. Answer the questions.**

a. 45% of the windows in an apartment building were dirty. There were 300 windows in the building.
   1. How many were dirty?
   2. How many were clean?

b. The cow's weight was .8 of the bull's weight. The bull weighed 2000 pounds. How much heavier was the bull than the cow?

c. Mary went 1.3 the distance that Ann went. Ann went 40 miles.
   1. How far did Mary go?
   2. How much farther did Mary go than Ann went?

d. 75% of the house is painted. The rest isn't painted. It will take 300 hours to paint the entire house.
   1. How long will it take to paint the part of the house that is not painted?
   2. How long did it take to paint the part that is painted?

Part 6 **For each item, write the equations. Figure out the mystery number.**

a. You start with some number and add 11. Then you multiply by 3. Then you add 33. Then you divide by 2. You end up with 33. What number did you start with?

b. You start with some number and divide by 2. Then you divide by 3. Then you add 50. You end up with 95. What number did you start with?

Part 7 **Make a table and fill in all the missing numbers. Then answer each question.**

> There are boys and girls in Williams School and Jefferson School.

Facts

1. 217 students attend Williams School.
2. 137 girls attend Jefferson School.
3. 129 boys attend Williams School.
4. At Jefferson School, there are 29 fewer boys than girls.

Items

a. At Williams School, there are 59 fewer teachers than girls. How many teachers are there at Williams School?

b. The number of students in Davis School is 184 more than the number of students in Williams School. How many students are in Davis School?

c. How many students are in Jefferson School?

Part J

a. $b \times h = A$; $12 \times 11 = A$; A = 132 sq ft

b. $b \times h = A$; $15\left(\frac{105}{15}\right) = 105$; h = 7 in

c. $b \times h = A$; $b \times 10 = 62$; $b = 6\frac{2}{10}$ in

d. $b \times h = A$; $17 \times 20 = A$; A = 340 sq m

Part K

a. $\frac{\text{tablets}}{\text{boxes}}$ $\frac{8}{1}\left(\frac{\frac{512}{8}}{\frac{512}{8}}\right) = \frac{512}{\frac{512}{8}}$; 64 boxes

b. $\frac{\text{tubs}}{\text{gal}}$ $\frac{1}{61}\left(\frac{\frac{549}{61}}{\frac{549}{61}}\right) = \frac{\frac{549}{61}}{549}$; 9 tubs

c. $\frac{\text{engines}}{\text{cylinders}}$ $\frac{1}{6}\left(\frac{\frac{234}{6}}{\frac{234}{6}}\right) = \frac{\frac{234}{6}}{234}$; 39 engines

d. $\frac{\text{sec}}{\text{buttons}}$ $\frac{5}{6}\left(\frac{\frac{1}{6}}{\frac{1}{6}}\right) = \frac{\frac{5}{6}}{1}$; $\frac{5}{6}$ sec

e. $\frac{\text{mi}}{\text{hr}}$ $\frac{580}{7}\left(\frac{\frac{1}{7}}{\frac{1}{7}}\right) = \frac{\frac{580}{7}}{1}$; $82\frac{6}{7}$ mi

Independent Work

Make the ratio tables and answer the questions in part 5.
Figure out the mystery number for each problem in part 6.

Part 8 **Work each problem. Make a ratio table if you need one.**

a. $\frac{5}{7}$ of the students wore jackets during recess. 46 students did not wear jackets.
   1. How many students were there?
   2. How many wore jackets?

b. At Harris school, $\frac{8}{10}$ of the jackets were old. There were 32 jackets that were new.
   1. How many jackets were old?
   2. How many jackets were there in all?

c. On a trip, there were 3 cold days for every 4 warm days. If there were 12 cold days, how many warm days were there?

d. A train went 4 miles for every 3 miles a car went. The car went 198 miles. How far did the train go?

e. A bike traveled 12 feet per second. How long did it take the bike to go 132 feet?

f. The dog weighed $\frac{3}{10}$ as much as the pig. The pig weighed 360 pounds. How much did the dog weigh?

g. The ratio of sand to dirt in a planting mixture is 2 to 7. If the mixture has 600 pounds of sand, how many pounds of dirt does it have?

h. A car travels at the rate of 389 miles in 4 hours. How many miles per hour does the car travel?

i The regular price of a bed was 1.3 times the sales price. The regular price was $104.
   1. What was the sale price?
   2. How much more was the regular price than the sales price?

Part 9 **Answer the questions. If necessary round your answer to hundredths.**

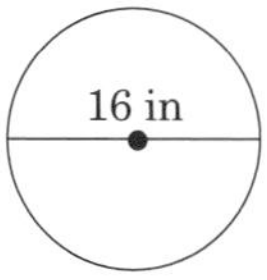

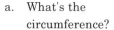

a. What's the circumference?

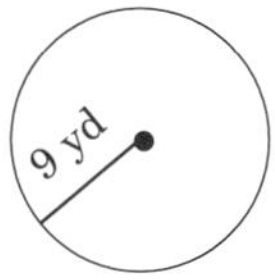

b. What's the circumference?

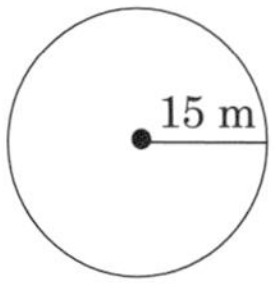

c. What's the diameter?

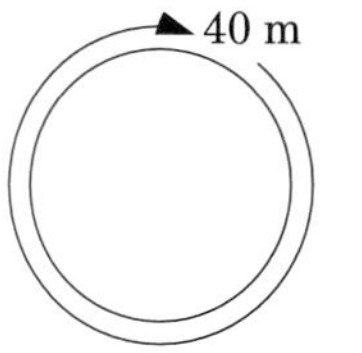

d. What's the radius?

e. What is the area of the triangle?

f. What is the perimeter of the triangle?

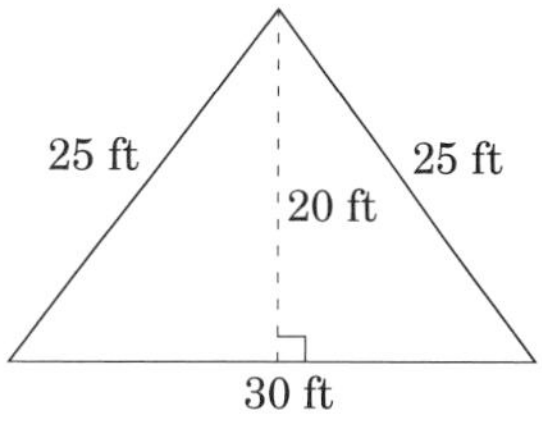

Lesson 69

Part 1

Work each item. If necessary, round your answer to hundredths.

a.

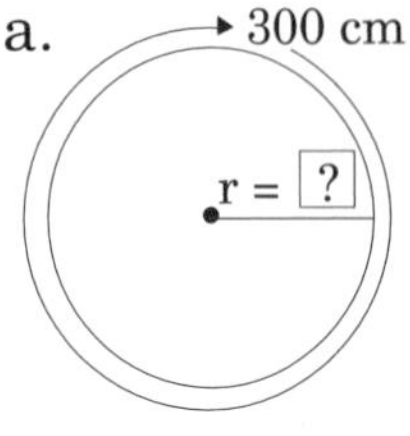

b.

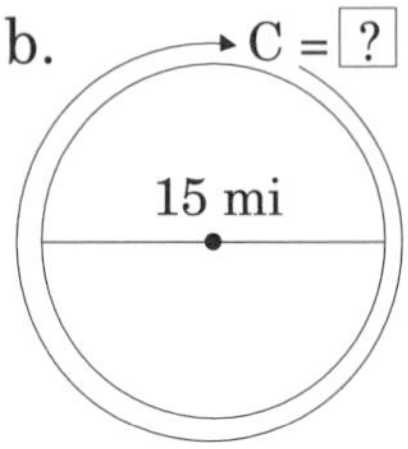

c.

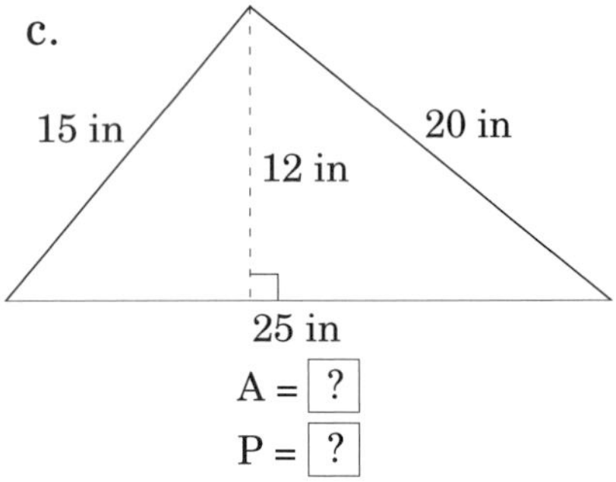

d.

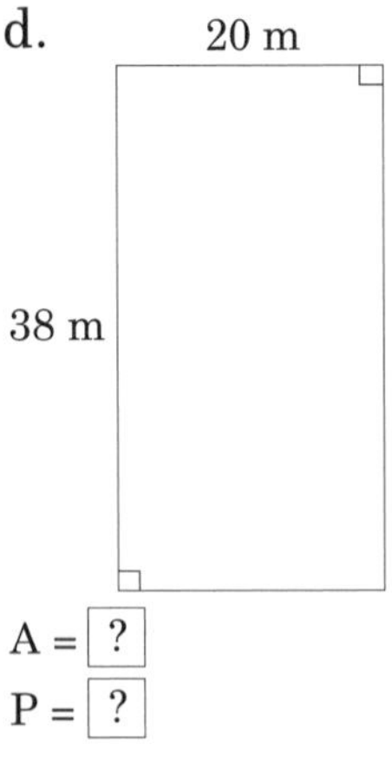

e.

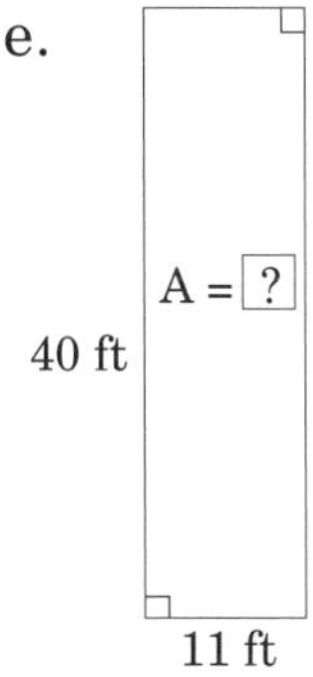

f.

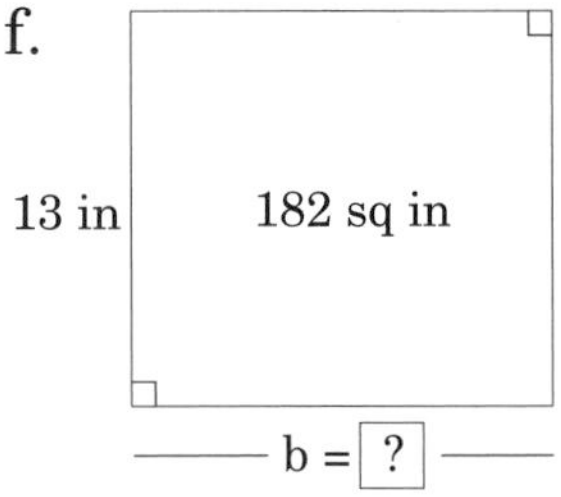

Part 2

Use an inequality sign or an equal sign to answer each question.

a. Which is larger, .8 or $\frac{17}{20}$?

b. Which is larger, $\frac{9}{25}$ or .34?

c. Which is larger, $\frac{7}{20}$ or .41?

d. Which is larger, $\frac{3}{2}$ or 1.5?

e. Which is larger, $\frac{3}{5}$ or .65?

Part 3 **For each item, write the equations. Figure out the mystery number.**

a. You start with some number. You subtract 2. Then you divide by 20. Then you subtract 11. You end up with 100. What's the mystery number?

b. You start with some number. You divide the number by 4. Then you subtract 10. Then you divide by 2. You end up with 1. What's the mystery number?

c. You start with some number. Then you add 56. Then you divide by 70. Then you subtract 1. You end up with 0. What's the mystery number?

Part 4 **Make a table and fill in all the missing numbers. Then answer each question.**

Two recreation parks have farm animals and wild animals. The parks are Animal World and Living Adventure.

Facts

1. There are 50 farm animals at Living Adventure.
2. The total number of animals at Animal World is 236.
3. At Living Adventure the number of wild animals is 149 more than the number of farm animals.
4. The total number of farm animals at both parks is 35 more than the number of farm animals at Living Adventure.

Items

a. In which park are there more farm animals?

b. What's the total number of animals at Living Adventure?

c. The number 400 tells about the total for what kind of animals?

d. What's the total number of farm animals?

e. How many more animals are in Living Adventure park than Animal World?

Part 5

Work each item. Show the answer as a whole number or mixed number, if possible.

a. $1 \times 3 \times \frac{12}{10} =$ [redacted]

b. $\frac{2}{7} \times \frac{5}{3} \times 2 =$ [redacted]

c. $\frac{15}{3} - 2 =$ [redacted]

d. $\frac{1}{8} + 4 + \frac{3}{5} =$ [redacted]

Part 6

For each item, change the percent or decimal value into a fraction. Work each problem.

a. The cost of the refrigerator was 1.5 times the cost of the TV. The refrigerator cost $705.

1. How much did the TV cost?
2. What's the difference in the price of the two products?

b. The price of a calculator is 5% the price of a stereo. The stereo costs $440.

1. How much less is the price of the calculator?
2. How much does the calculator cost?

c. .08 of the students in Briggs School had worked on a farm. There were 16 students in Briggs School who had worked on a farm.

1. How many total students were in Briggs School?
2. How many students had **not** worked on a farm?

d. A car traveled at 1.4 the speed of a bus. The car traveled at 56 miles per hour.

1. What was the speed of the bus?
2. What was the difference in the speed of the two vehicles?

Independent Work

Make the ratio tables and answer the questions in part 6.

Part 7

Copy and complete the table to show the x and y values for each point.

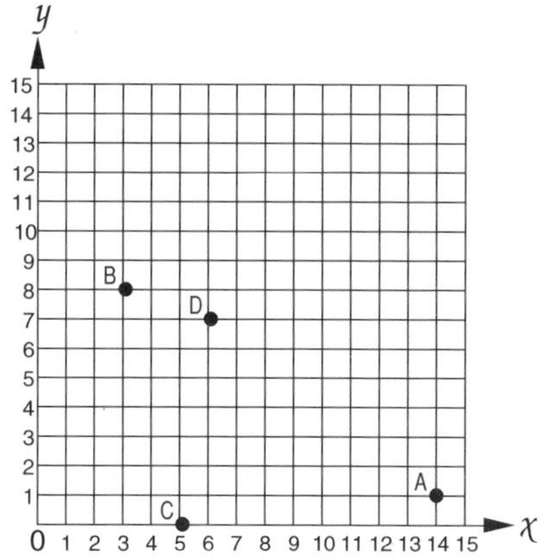

	x	y
A	■	■
B	■	■
C	■	■
D	■	■

Part 8

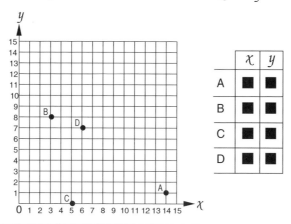

For some problems, write a ratio equation or ratio table. For other problems, make a number family. Answer the questions.

a. $\frac{4}{9}$ of the corn was ripe. What fraction of the corn was not ripe?

b. James picked 19 quarts of berries in 2 hours. How many quarts of berries per hour did James pick?

c. Men and women worked in a building. There were 57 men in the building. How many women worked in the building if 123 people worked in the building?

d. Men and women worked in a building. There were 4 men for every 7 women. How many men worked in the building if there were 91 women?

e. Men and women worked in a building. There were 57 more men than women. 123 women worked in the building. How many men worked in the building?

f. Men and women worked in a building. 4 out of every 7 people were men. 91 people worked in the building.
   1. How many were men?
   2. How many were women?

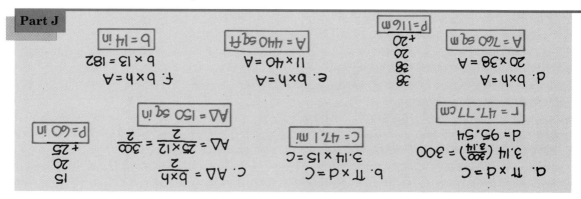

Part J

a. $\pi \times d = C$
$3.14 \left(\frac{300}{3.14}\right) = 300$
$d = 95.54$
r = 47.77 cm

b. $\pi \times d = C$
$3.14 \times 15 = C$
C = 47.1 mi

c. $A\Delta = \frac{b \times h}{2}$
$A\Delta = \frac{25 \times 12}{2} = \frac{300}{2}$
AΔ = 150 sq in

15
20
+25
P=60 in

d. $b \times h = A$
$20 \times 38 = A$
A = 760 sq m

38
38
20
+20
P=116 m

e. $b \times h = A$
$11 \times 40 = A$
A = 440 sq ft

f. $b \times h = A$
$b \times 13 = 182$
b = 14 in

Part K

a. $\frac{1}{1} \times \frac{3}{1} \times \frac{12}{10} = \frac{36}{10} = \boxed{3\frac{6}{10}}$

b. $\frac{2}{7} \times \frac{5}{3} \times \frac{2}{1} = \boxed{\frac{20}{21}}$

c.
$$\begin{array}{rcr} \frac{15}{3} & = & \frac{15}{3} \\ -2 & = & -\frac{6}{3} \\ \hline & & \frac{9}{3} = \boxed{3} \end{array}$$

d.
$$\begin{array}{rcr} \frac{1}{8}\left(\frac{5}{5}\right) & = & \frac{5}{40} \\ 4 & = & \frac{160}{40} \\ +\frac{3}{5}\left(\frac{8}{8}\right) & = & +\frac{24}{40} \\ \hline & & \frac{189}{40} = \boxed{4\frac{29}{40}} \end{array}$$

If you start with the mystery number, go up 120 feet and then come back down 135 feet, how many fish will you see?

Test 7

Part 1

Use an inequality sign to answer each question.

a. Which is more, $\frac{2}{3}$ or $\frac{16}{21}$?

b. Which is more, .7 or $\frac{24}{30}$?

Part 2

For each item, make a ratio table or ratio equation and answer the questions. Change the percent or decimal values into fractions.

a. The regular price of a TV was 125% of the sale price. The difference in price was $200.

   1. What was the regular price of the TV?
   2. What was the sale price?

b. Mr. James cut boards so they were .7 of their original length. The original length of the boards was 80 centimeters.

   1. How many centimeters did he cut from each board?
   2. How long was each board after it was cut?

c. A recipe uses 16 ounces of flour for every 5 eggs. If a cook uses 3 eggs, how many ounces of flour are needed?

d. A mill makes 12 pounds of sawdust per second. How long will it take the mill to produce 100 pounds of sawdust?

e. 48 identical woodblocks weigh 17 pounds. How many woodblocks would weigh 1 pound?

Part 3

For each item, write the equations. Figure out the mystery number.

a. You start with a number and triple it. Then you subtract 100. Then you divide by 8. You end up with 1. What number did you start with?

b. You start with the mystery number and multiply it by 7. Then you subtract 10. Then you add 50. You end up with 61. What's the mystery number?

Part 4

For each item, show whether M is more or less than 1. Then show what M equals.

a. 4M = 3

M [redacted] 1

M = [redacted]

b. 16M = 17

M [redacted] 1

M = [redacted]

Part 5

Work each item.

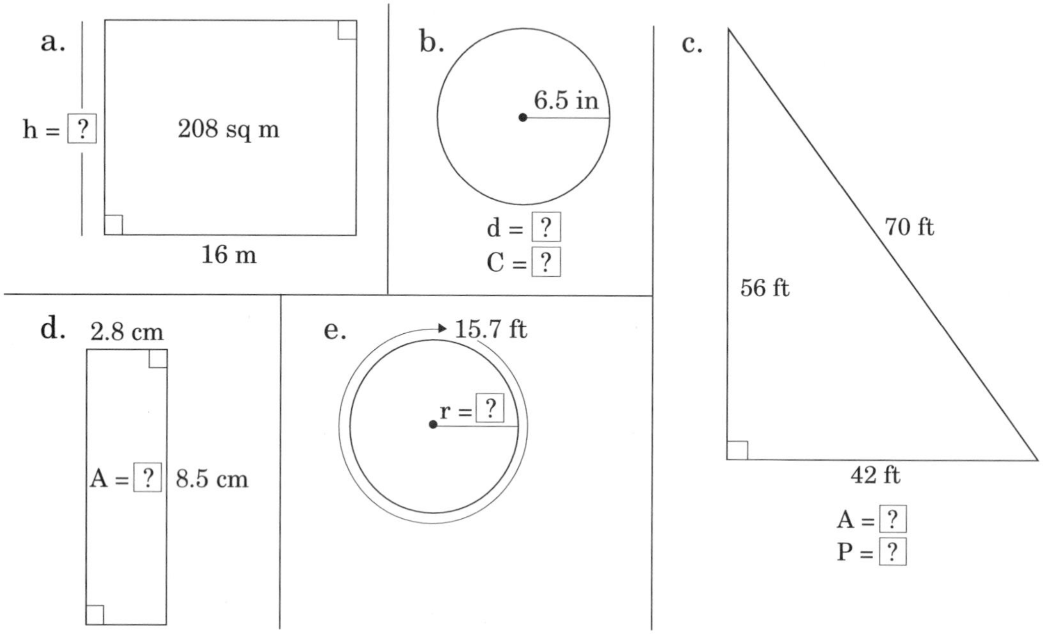

Part 6

Round each decimal value to hundredths.

a. 3.416 b. 28.0316 c. 29.008

Final Test

Final Test

Part 1 **Copy the table. For each row, write the division problem and the answer, the fraction and the mixed number.**

	Division	Fraction	Mixed number
a.		$\frac{5}{4}$	
b.			$2\frac{1}{3}$
c.	$9\overline{)65}$		

Part 2 **Copy each equation and write the missing number.**

a. $\blacksquare \div 7 = 9$

b. $\blacksquare + 256 = 308$

Part 3 **Work each problem. If an answer is more than 1, write it as a mixed number.**

a. $2 + \frac{1}{3} + \frac{4}{5} = \blacksquare$

b. $\frac{3}{5} \times \frac{1}{3} = \blacksquare$

c. $\frac{7}{8} - \frac{1}{2} = \blacksquare$

d. $\frac{3}{5} \times 5 \times \frac{7}{4} = \blacksquare$

e. $\frac{3}{8} = \frac{24}{\blacksquare}$

f. $4 = \frac{\blacksquare}{8}$

Part 4 **Copy and complete each equation.**

a. $\frac{3}{4} = \frac{10}{\blacksquare}$

b. $\frac{11}{7} = \frac{\blacksquare}{2}$

Part 5 **Copy each item. Show each answer as a fraction.**

a. $3f = 19$

$f = \blacksquare$

b. $5R = 4$

$R = \blacksquare$

Part 6 **Simplify each value.**

a. $\frac{6}{36} = \blacksquare$

b. $\frac{8}{12} = \blacksquare$

c. $\frac{50}{65} = \blacksquare$

Part 7 **For each item, use the inequality sign to show if R is more than 1 or less than 1.**

a. $17R = 16$

R ■ 1

b. $5R = 9$

R ■ 1

Part 8 **Use an inequality sign to answer each question.**

a. Which is more, .23 or .203?

b. Which is more, $\frac{3}{4}$ or $\frac{8}{12}$?

c. Which is more, .6 or $\frac{19}{30}$?

Part 9 **Copy and complete the table.**

		$\frac{■}{100}$	Hundredths decimal	%
a.	$\frac{3}{25}$			
b.	$\frac{10}{4}$			
c.	$\frac{220}{50}$			

Part 10 **Make number families. Answer the questions.**

a. $\frac{2}{5}$ of the workers took the bus to work. What fraction of the workers did not take the bus to work?

b. $\frac{14}{52}$ of the weeks had temperatures below freezing. What fraction of the weeks had temperature above freezing?

c. A swamp has 78 birds in it. 18 of those birds are geese. What fraction of the birds are not geese?

Part 11 Work each problem and answer each question.

a. 4 cartons weigh 6 pounds. How many pounds do 11 cartons weigh?

b. If there are 3 grams of salt in each gallon of water, how many gallons of water are needed for 20 grams of salt?

c. The regular price of a bed is 120% of the sale price. A person who bought the bed on sale would save $60.

1. How much is the regular price of the bed?
2. How much is the sale price?

d. $\frac{3}{7}$ of the students at Ellis School are boys. There are 224 girls in Ellis School.

1. How many total students are in Ellis school?
2. How many students are boys?

Part 12 Write equations for each problem. Answer the questions.

a. You start with a number and divide it by 2. Then you subtract 50 and end up with 1. What number did you start with?

b. You start with the mystery number and multiply it by 6. Then you subtract 81. You end up with 21. What's the mystery number?

Part 13 **Work the problem and answer the questions.**

> There are geese and other birds on two wildlife preserves. The preserves are Blue Point Preserve and Gill Island.

Facts

1. The total number of birds on Gill Island is 421.
2. The total number of geese for both preserves is 402.
3. There are 184 geese on Blue Point Preserve.
4. There are 51 fewer geese than other birds on Blue Point Preserve.

Items

a. How many birds that are not geese are on Blue Point Preserve?

b. How many birds are on Blue Point Preserve?

c. How many geese are on Gill Island?

d. What's the total number of birds that are not geese?

Part 14 **Work each problem. Show the answer as a mixed number.**

a. $76\overline{)629}$ b. $43\overline{)232}$

Part 15 **Work each item.**

a. How many degrees are in a full circle?

b. How many degrees are in a corner of a rectangle?

c. How many degrees are in a half circle?

d. When two lines are perpendicular they form an angle that is ■ degrees.

Part 16

Two of the lines are parallel. Complete the equations to show the number of degrees for each angle.

$\angle r = 78°$

a. $\angle t =$ ■

b. $\angle v =$ ■

c. $\angle n =$ ■

d. $\angle w =$ ■

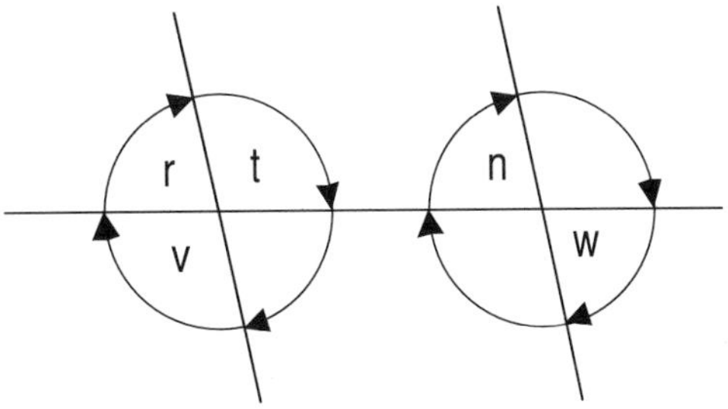

Part 17

Figure out the height of the rectangle.

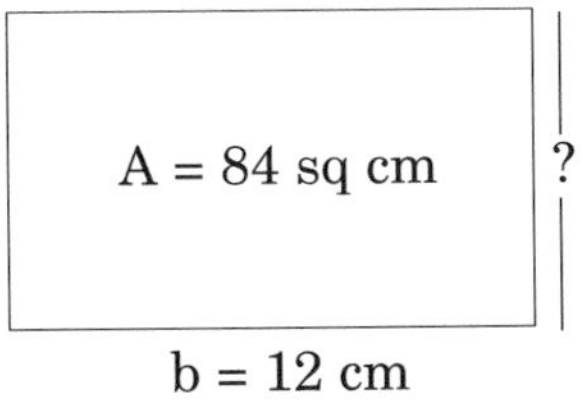

Part 18

Write the x and y values for each point.

A $x =$ ■, $y =$ ■

B $x =$ ■, $y =$ ■

C $x =$ ■, $y =$ ■

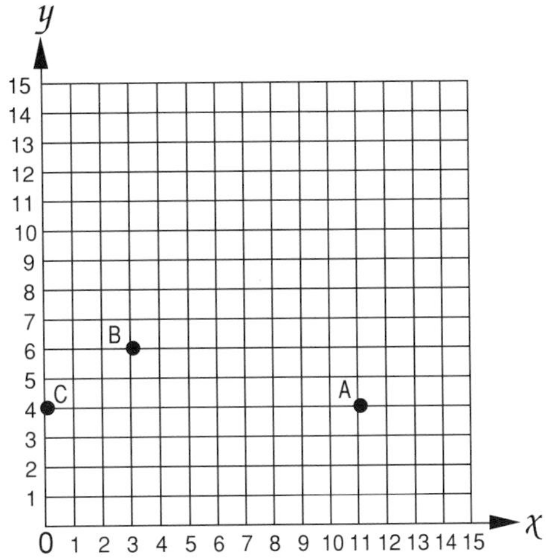

Part 19

Work each item. Do not use a calculator for items a through d.

a. 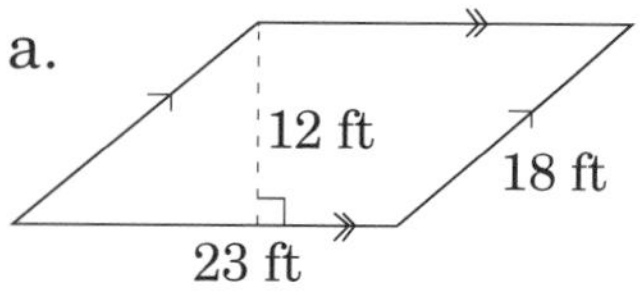

What is the perimeter?
What is the area?

b. 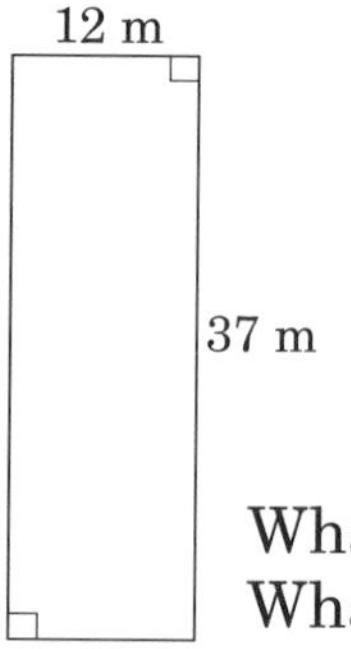

What is the perimeter?
What is the area?

c. 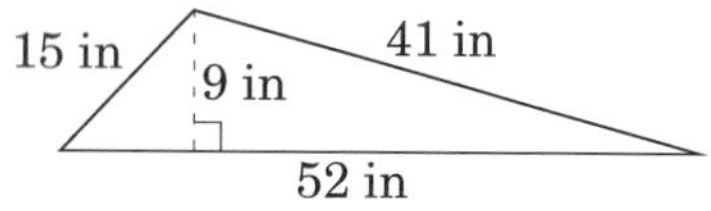

What is the perimeter?
What is the area?

d. 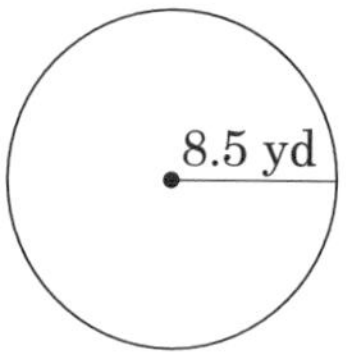

What is the diameter?
What is the circumference?

e. 

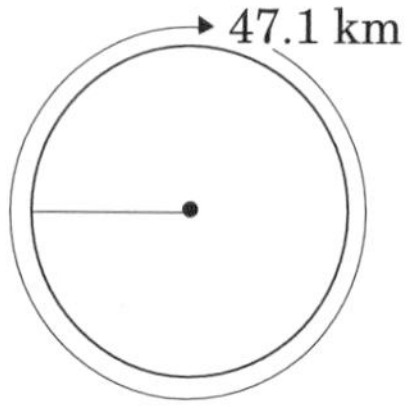

What is the diameter?
What is the radius?

Part 20

Use your calculator to work each problem. Show the answer as a mixed number.

a. $600 \div 80 =$ [redacted]

b. $15 \div 12 =$ [redacted]

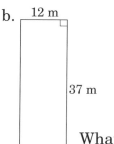

Answer Key

31

Part 8
a. A = 160 sq km
c. A = 320 sq cm

Part 9
a. P = 68 ft
c. P = 160 in

Part 10
a. 81 lb
c. 98 dogs

Part 11

a.

2	5
22	9
7	6

c.

7	71
4	62
20	110
5	65

Part 12
a. 162 sq yd [of carpet]
c. 189 canoes

Part 13
a. $\frac{2}{3}\left(\frac{8}{8}\right) = \frac{16}{24}$; $\frac{2}{3} = \frac{16}{24}$
c. $\frac{2}{5}\left(\frac{6}{6}\right) = \frac{12}{30}$; $\frac{2}{5} = \frac{12}{30}$
e. $\frac{1}{7}\left(\frac{30}{30}\right) = \frac{30}{210}$; $\frac{1}{7} = \frac{30}{210}$

Part 14

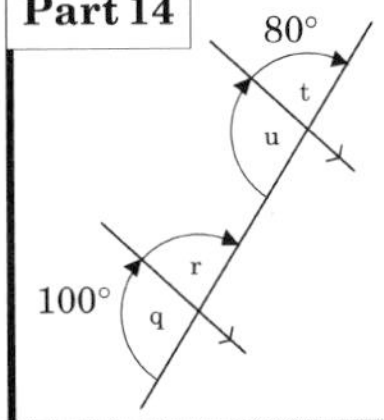

Part 15

a. 27 (4) = 108	$27\overline{)108}$ = 4	$\frac{108}{27} = 4$
c. 9 (31) = 279	$9\overline{)279}$ = 31	$\frac{279}{9} = 31$

32

Part 10
a. 74 lb
c. vegetables that are not potatoes, Jackson School

Part 11
a. $\frac{3}{4}\left(\frac{21}{22}\right) = \frac{63}{88}$
c. $\frac{3}{7} - \frac{3}{7} = \frac{0}{7}$
e. $\frac{8}{5} + \frac{6}{5} - \frac{2}{5} = \frac{12}{5}$
g. $\frac{16}{9} - \frac{0}{9} = \frac{16}{9}$
i. 1 x 15 = 15
k. 0 x 264 = 0
m. $\frac{7}{3} + \frac{3}{3} = \frac{10}{3}$

Part 12

a.

10	99
19	9
0	199
1	189

Part 13
a. 90°
c. 360°

Part 14
a. $\frac{3}{5}\left(\frac{9}{9}\right) = \frac{27}{45}$; $\frac{3}{5} = \frac{27}{45}$
c. $\frac{12}{11}\left(\frac{5}{5}\right) = \frac{60}{55}$; $\frac{12}{11} = \frac{60}{55}$

Part 15
a. 980 workers
c. 189 hawks

Part 16

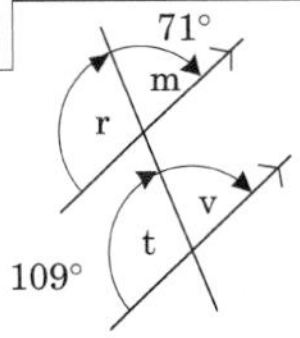

33

Part 10
a. 36 people
c. Ms. Tyler's class

Part 11
a. $2 = \frac{56}{28}$
c. $\frac{28}{7} = 4$
e. $46 = \frac{230}{5}$

Part 12
a. 75 green cubes
c. 81 bushels

Part 13

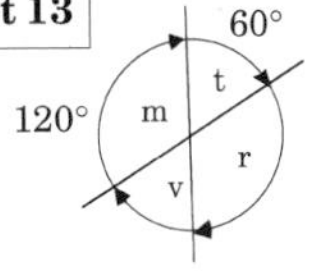

Part 14
a. $\frac{3}{8}\left(\frac{4}{4}\right) = \frac{12}{32}$; $\frac{3}{8} = \frac{12}{32}$
c. $\frac{1}{9}\left(\frac{10}{9}\right) = \frac{10}{81}$

Part 15
a. N = 208
c. J = 8237

Part 16

a.	$7\overline{)388}$ = $55\frac{3}{7}$	$\frac{388}{7} = 55\frac{3}{7}$
c.	$8\overline{)300}$ = $37\frac{4}{8}$	$\frac{300}{8} = 37\frac{4}{8}$

34

Part 8
a. 1628 lb
c. 124 in

Part 9

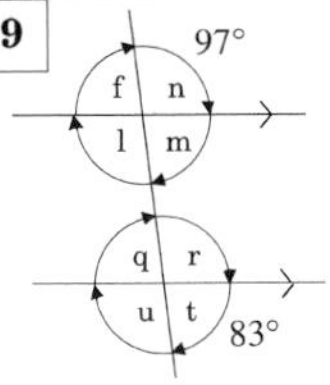

Part 10
a. 116 vehicles
c. 241 mi

Part 11
a. A = 192 sq in
P = 56 in
c. A = 2691 sq ft
P = 220 ft

Part 12

a.

0	10
6	22
10	30

c.

50	149
2	5
7	20

Part 13
a. $48
c. 130 piles

Part 14
a. $\frac{5}{6} \times \frac{7}{6} = \frac{35}{36}$
c. $\frac{13}{4} - \frac{8}{4} - \frac{5}{4} = \frac{0}{4}$
e. $\frac{2}{5}\left(\frac{40}{40}\right) = \frac{80}{200}$
g. $\frac{5}{3} + \frac{1}{3} - \frac{2}{3} = \frac{4}{3}$

35

Part 7

a. 202 people

c. Big Ride

Part 8

a. 243 men

Part 9

a. 556 books

Part 10

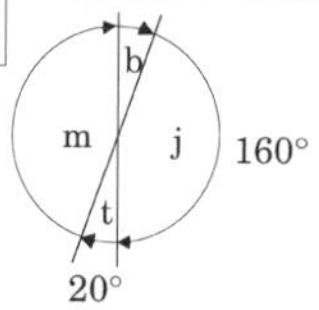

Part 11

a. 198 cats

c. 720 min

Part 12

a. T = 441

c. M = 71,310

Part 13

a. $\frac{286}{5} = 57\frac{1}{5}$ | $5\overline{)286}$ = $57\frac{1}{5}$

c. $\frac{478}{4} = 119\frac{2}{4}$ | $4\overline{)478}$ = $119\frac{2}{4}$

36

Part 1

b.

clean	2	144
dirty	8	576
shoes	10	720

d.

closed	3	276
open	1	92
doors	4	368

Part 7

b. A□ = 30 sq ft
P = 26 ft

Part 8

b. 240 ft

Part 9

b. $\frac{732}{5} = 146\frac{2}{5}$

Part 10

b. $8\overline{)224}$ = 28 | 8 (28) = 224 | $\frac{224}{8} = 28$

Part 11

b. $\frac{4}{5}$ —— $\frac{1}{5}$ ▶ $\frac{5}{5}$

d. $\frac{2}{9}$ —— $\frac{7}{9}$ ▶ $\frac{9}{9}$

Part 12

b. $\frac{7}{1}\left(\frac{30}{30}\right) = \frac{210}{30}$

$\frac{7}{1} = \frac{210}{30}$

d. $\frac{2}{9}\left(\frac{11}{11}\right) = \frac{22}{99}$

$\frac{2}{9} = \frac{22}{99}$

37

Part 8

b. $6\frac{72}{81}$

Part 9

b. A△ = 300 sq ft

Part 10

b.

good	11	132
flat	1	12
tires	12	144

Part 11

b.

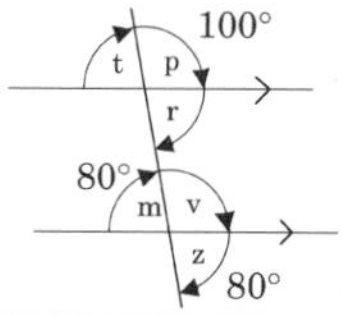

Part 12

b. 3597 lb

d. 9 cars [did not need repairs]

38

Part 8

b. P = 68 m

Part 9

b. 129 cities

d. 320 ripe watermelons

Part 10

b. $3\frac{82}{92}$

Part 11

b.

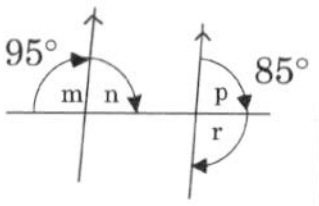

Part 12

b. $\frac{2}{5}$ —— $\frac{3}{5}$ ▶ $\frac{5}{5}$

Part 13

b.

girls	3	90
boys	4	120
children	7	210

Part 14

b.

8	8
23	11
3	7

39

Part 8

rubber — no rubber — shoes

b. $\frac{3}{10}$ —— $\frac{7}{10}$ ▶ $\frac{10}{10}$

Part 9

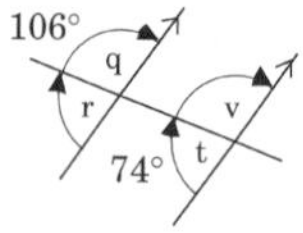

Part 10

b. $\frac{8}{3} - \frac{2}{3} = \frac{6}{3}$

d. $\frac{13}{4}\left(\frac{1}{3}\right) = \frac{13}{12}$

f. $\frac{9}{10} \times \frac{6}{10} = \frac{54}{100}$

h. 75 – 74 = 1

j. $\frac{1}{5} + \frac{4}{5} = \frac{5}{5}$

Part 11

b. $8\frac{50}{73}$

Part 12

b. 30 g

d. 240 g

Part 13

b. 13 (6) = 78 | $13\overline{)78}$ = 6 | $\frac{78}{13} = 6$

Part 14

b.

short-tailed	13	39
long-tailed	2	6
dogs	15	45

41

Part 6

b. $1\frac{1}{5}$

d. $8\frac{2}{6}$

f. $5\frac{5}{7}$

Part 9

b. 64 trout

Part 10

b.	$7\overline{)347}$ = $49\frac{4}{7}$	$\frac{347}{7} = 49\frac{4}{7}$
d.	$3\overline{)306}$ = 102	$\frac{306}{3} = 102$

Part 11

b. 1216 lb

d. 82 cookies

Part 12

b. $5\frac{8}{46}$ d. $4\frac{2}{28}$

Part 13

b.

0	10
20	110
1	15
3	25

Part 14

b. 16 bricks

Part 15

no radials, radials, tires

b. $\frac{7}{15}$ $\boxed{\frac{8}{15}}$ → $\frac{15}{15}$

42

Part 9

b. 150 cows

Part 10

b. $7\frac{8}{23}$

Part 11

b. A △ = 108 sq ft
P = 54 ft

Part 12

∠ z = 50°

∠ y = 130°

∠ t = 50°

Part 13

b. P = 547

Part 14

b. $\frac{197}{9} = 21\frac{8}{9}$

Part 15

b. $31\frac{4}{10} = \frac{314}{10}$

43

Part 7

b. 20 children

d. 406 peanuts

Part 8

not raining, raining, time

b. $\frac{3}{12}$ $\boxed{\frac{9}{12}}$ → $\frac{12}{12}$

Part 9

b. $8\frac{4}{38}$

Part 10

b. ∠ f = 65°

∠ m = 115°

∠ s = 65°

Part 11

b. $3 = \frac{150}{50}$

d. $\frac{1}{7}\left(\frac{100}{100}\right) = \frac{100}{700}$

f. $\frac{1}{8}$

h. $\frac{1}{8} \times \frac{20}{5} = \frac{20}{40}$

Part 12

b. A △ = 150 sq in
P = 60 in

44

Part 6

b. 137

d. 1359

Part 7

b. Roger Farm

Part 8

b. A □ = 192 sq in
P = 72 in

Part 9

b. $\frac{9}{21}$

d. $\frac{27}{25}$

f. $\frac{6}{2}$

Part 10

b. ∠ e = 63°

∠ b = 117°

∠ c = 63°

Part 13

b.

6	2
25	40
12	14
5	0

Part 14

b.

7	105
4	60
11	165

45

Part 7

b. 210 sneakers

Part 8

b. $690\frac{1}{2}$

d. $3\frac{19}{78}$

Part 9

b. ∠ s = 132°

∠ n = 132°

∠ t = 48°

Part 10

b. 18 hr

Part 11

b.	2 (28) = 56	$2\overline{)56}$ = 28	$\frac{56}{2} = 28$

Part 12

b.

1	14
2	25
3	36

Part 13

b. 666 doors

d. $247

46

Part 6

b. 1. 105 oz

2. 385 oz

Part 7

b. $3\frac{3}{92}$

Part 8

b. $\frac{63}{9} = 7$

d. $\frac{7}{9}\left(\frac{4}{4}\right) = \frac{28}{36}$

f. $\frac{20}{6}$

h. $27\frac{1}{3}$

Part 9

dirty, clean, sweaters

b. $\frac{2}{9}$ $\boxed{\frac{7}{9}}$ → $\frac{9}{9}$

Part 10

b. 72 years old

Part 11

b. ∠ h = 68°

∠ g = 112°

∠ j = 68°

47

Part 8

b. 1. 36 cards

2. 16 cards

Part 9

b. $40 \quad \begin{array}{r} 65 \\ -\ 27 \\ \hline 38 \end{array}$ d. $7000 \quad \begin{array}{r} 5076 \\ +\ 2324 \\ \hline 7400 \end{array}$

Part 10

b. 90 gal

Part 11

b. A △ = 192 sq in
P = 72 in

Part 12

b. 35 = 5 x 7

Part 13

∠ d = 43°

∠ m = 137°

48

Part 8

b. 1. 60 worms

2. 510 animals

Part 9

b. $8\frac{10}{36}$

Part 10

b.

1	72
4	108
0	60

Part 11

b. $1053

d. 708 students

f. not work $\frac{2}{7}$ — work $\boxed{\frac{5}{7}}$ → days $\frac{7}{7}$

Part 12

b. $7\frac{1}{9}$

d. $43\frac{3}{4}$

f. $\frac{3}{8}\left(\frac{9}{9}\right) = \frac{27}{72}$

Part 13

b. 33 = 3 x 11

d. 50 = 2 x 5 x 5

49

Part 8

b. 1472 buttons

Part 9

b. 20 = 2 x 2 x 5

d. 120 = 2 x 2 x 2 x 3 x 5

Part 10

b.

			total
	88	99	187
	134	66	200
total	222	165	387

Part 11

b.

$6\overline{)120}$ = 20	$\frac{120}{6} = 20$	6 (20) = 120

Part 12

b. intersecting

d. parallel

Part 13

b. $245

d. 2344 animals

Part 14

b. difficult $\frac{2}{10}$ — not difficult $\boxed{\frac{8}{10}}$ → questions $\frac{10}{10}$

Part 15

b. $57\left(\frac{19}{57}\right) = 19$

d. $16\left(\frac{4}{16}\right) = 4$

51

Part 8

b. 1. 255 trees

2. 225 trees

Part 9

b. $9\frac{20}{24}$

Part 10

b. poor $\frac{4}{79}$ — good $\boxed{\frac{75}{79}}$ → tires $\frac{79}{79}$

Part 11

b. $\frac{13}{5}$

d. $\frac{4}{8}$

f. $14\left(\frac{35}{14}\right) = 35$

Part 12

b. P = 60 in
A △ = 150 sq in

Part 13

b. $\frac{3}{100} = .03$

d. $\frac{405}{10} = 40.5$

Part 14

b.

8 (73) = 584	$8\overline{)584}$ = 73	$\frac{584}{8} = 73$

52

Part 9

b. red $\boxed{\frac{5}{9}}$ — not red $\frac{4}{9}$ → apples $\frac{9}{9}$

Part 10

b. 1. 160 frogs

2. 220 frogs

Part 11

∠ b = 135°

∠ f = 135°

Part 12

b.	.074	$\frac{74}{1000}$
d.	.054	$\frac{54}{1000}$

Part 13

	x	y
B	6	5
D	0	6

Part 14

50	0
30	80
1	196
0	200

Part 15

b. 3

d. $7\frac{16}{53}$

Part 16

b. 497 10th graders

d. 539 students

53

Part 9

b. 1. 100 apples
2. 20 apples

Part 10

b. $R = \frac{8}{5} = 1\frac{3}{5}$

d. $V = \frac{54}{15} = 3\frac{9}{15}$

Part 11

b. $\frac{1}{5}$

d. $\frac{10}{10}$

Part 12

b. $\frac{2}{3}\left(\frac{33}{33}\right) = \frac{66}{99}$

$\frac{2}{3} = \frac{66}{99}$

d. $\frac{7}{4}\left(\frac{4}{3}\right) = \frac{28}{12}$

Part 13

10	10
8	9
20	15
2	6
0	5

Part 14

b. $12\left(\frac{100}{12}\right) = 100$ | $12\overline{)100}$ = $8\frac{4}{12}$

54

Part 9

b. $2\frac{44}{52}$

d. $\frac{9}{7}\left(\frac{4}{4}\right) = \frac{36}{28}$

Part 10

b. 1. 360 girls
2. 900 students

Part 11

b. 380 tons

d. gold

Part 12

b. $8\left(\frac{1}{8}\right) = 1$

d. $15\left(\frac{37}{15}\right) = 37$

Part 13

b. A △ = 65 sq ft
P = 42 ft

Part 14

b. $2\frac{7}{56}$

55

Part 10

b. black $\frac{4}{5}$, not black $\frac{1}{5}$ → cards $\frac{5}{5}$

d. dif $\frac{3}{5}$, baskets $\frac{5}{5}$ → cards $\frac{8}{5}$

f. sick $\frac{3}{11}$, not sick $\frac{8}{11}$ → children $\frac{11}{11}$

Part 11

∠ k = 30°

∠ m = 150°

Part 12

b. 82 sec

d. 150 boys

Part 13

b.

48	32
120	20
12	38

Part 14

b. J = 35

Part 15

b. $\frac{3}{5}\left(\frac{8}{8}\right) = \frac{24}{40}$

d. $11 = \frac{396}{36}$

Part 16

b. $17\left(\frac{190}{17}\right) = 190$

d. $18\left(\frac{19}{18}\right) = 19$

56

Part 6

b. $\frac{50}{65} = \frac{2 \times 5 \times \not{5}}{\not{5} \times 13} = \frac{10}{13}$

d. $\frac{24}{48} = \frac{\not{2} \times \not{2} \times \not{2} \times \not{3}}{\not{2} \times \not{2} \times \not{2} \times 2 \times \not{3}} = \frac{1}{2}$

Part 7

b. $12\left(\frac{1}{12}\right) = 1$

d. $40\left(\frac{96}{40}\right) = 96$

Part 8

b. 8 (141) = 1128 | $\frac{1128}{8} = 141$ | $8\overline{)1128}$ = 141

Part 9

A □ = 600 sq ft
P = 110 ft

Part 10

b. 69 days

d. 60 goats

Part 11

b.	$\frac{6}{100}$	.06
d.	$\frac{148}{100}$	1.48

Part 12

b. in $\frac{2}{9}$, not in $\frac{7}{9}$ → cows $\frac{9}{9}$

d. 852 children

Part 13

b. M = 69

d. $\frac{6}{7}\left(\frac{41}{41}\right) = \frac{246}{287}$

f. $15 = \frac{45}{3}$

57

Part 10

b. $60

Part 11

	x	y
B	1	7
D	6	6

Part 12

b. $\frac{36}{40} = \frac{\cancel{2} \times \cancel{2} \times 3 \times 3}{\cancel{2} \times \cancel{2} \times 2 \times 5} = \frac{9}{10}$

d. $\frac{30}{60} = \frac{\cancel{2} \times \cancel{3} \times \cancel{5}}{\cancel{2} \times 2 \times \cancel{3} \times \cancel{5}} = \frac{1}{2}$

Part 13

b. $R = \frac{10}{3}$

d. $Q = \frac{6}{96}$

Part 14

b. 4.3

d. 145.06

Part 15

b. $\frac{54}{9} = 6$

d. $3 = \frac{12}{4}$

Part 16

d. $\frac{580}{10} = 58$

g. $\frac{47}{3} = 15\frac{2}{3}$

58

Part 7

b.	$\frac{308}{100}$	3.08	$3\frac{8}{100}$
d.	$\frac{17,015}{1000}$	17.015	$17\frac{15}{1000}$

Part 8

b. 90 gal

Part 9

a.

58	92	150
302	203	505
360	295	655

b.

23	1	24
43	32	75
66	33	99

Part 10

b. $72.90

Part 11

b. $\frac{75}{10} = 7\frac{5}{10}$

Part 12

b. 1464 lb

Part 13

b. $\frac{17}{20}$

d. $\frac{49}{7} = 7$

f. $\frac{20}{11}\left(\frac{3}{3}\right) = \frac{60}{33}$

Part 14

b. $6\frac{9}{82}$

59

Part 8

b. 1. 400 hardback books

2. 320 paperback books

Part 9

b. C = 3.14 mi

Part 10

	x	y
B	7	3
D	8	1

Part 11

b. $8\overline{)296}$ = 37 | 8 (37) = 296

Part 12

b. 471 cards

Part 13

b. $18\left(\frac{1}{18}\right) = 1$

d. $J = \frac{7}{2} = 3\frac{1}{2}$

Part 14

	jacks	not jacks	cards
b.	$\frac{4}{52}$	$\frac{48}{52}$ →	$\frac{52}{52}$

Part 15

b. $\frac{15}{20} = \frac{3 \times \cancel{5}}{2 \times 2 \times \cancel{5}} = \frac{3}{4}$

d. $\frac{16}{20} = \frac{\cancel{2} \times \cancel{2} \times 2 \times 2}{\cancel{2} \times \cancel{2} \times 5} = \frac{4}{5}$

61

Part 10

	dirty	not dirty	windows
b.	$\frac{66}{84}$	$\frac{18}{84}$ →	$\frac{84}{84}$

Part 11

b. $\frac{17}{24}$

Part 12

b. $\frac{128}{8} = 16$

Part 13

b. 16 mi

Part 14

b. $36\frac{1}{2} = \frac{73}{2}$

d. $1\frac{3}{50} = \frac{53}{50}$

62

Part 2

b. 27 hr

Part 10

b. 1. $\frac{14}{19}$ 2. $\frac{5}{19}$

Part 11

b. A□ = 117 sq ft

P = 48 ft

Part 12

b. $\frac{9}{50}$

d. $\frac{120}{10} = 12$

f. $\frac{3}{9}\left(\frac{4}{4}\right) = \frac{12}{36}$

h. $\frac{3}{7}\left(\frac{7}{10}\right) = \frac{21}{70}$

Part 13

$\angle$ d = 72°

63

Part 6

b. 36 girls

Part 8

	squeak	no squeak	doors
b.	$\frac{3}{7}$	$\frac{4}{7}$	$\frac{7}{7}$

Part 9

b. 1. 60 people
2. 30 people

Part 10

b. $R = \frac{1}{12}$

d. $R = \frac{7}{7} = 1$

f. $J = \frac{8}{9}$

Part 11

	x	y
K	0	6
P	7	0

Part 12

b.	$\frac{1202}{100}$	12.02	$12\frac{2}{100}$
d.	$\frac{23}{10}$	2.3	$2\frac{3}{10}$

Part 13

b. A□ = 600 sq in
P = 110 in

Part 14

b. $\frac{3}{5}\left(\frac{5}{5}\right) = \frac{15}{25}$ d. $3 = \frac{42}{14}$ f. $18 = \frac{72}{4}$

64

Part 6

b. 36 secretaries

d. 1. 88 employees
2. 33 employees

Part 7

b. 1. $\frac{27}{27}$
2. $\frac{12}{27}$

Part 8

b. $\frac{29}{60}$

Part 9

d. $\frac{7}{12} < .6$

Part 10

b. $\frac{9}{32}$ d. $\frac{7}{5}$ f. $\frac{3}{5}\left(\frac{11}{10}\right) = \frac{33}{50}$

Part 11

	x	y
K	6	1
N	7	8

Part 12

b.	$\frac{8}{5}$	$\frac{160}{100}$	1.60	160%
d.	$\frac{2}{25}$	$\frac{8}{100}$	.08	8%

Part 13

$\angle$ p = 19°

65

Part 8

b. 42 babies

d. 1. \$36
2. \$96

Part 9

b. r = 8 mi

Part 10

b.	$\frac{2}{10}$	$\frac{20}{100}$	.20	20%
d.	$\frac{5}{2}$	$\frac{250}{100}$	2.50	250%

Part 11

b. $\frac{12}{14} = \frac{\cancel{2} \times 2 \times 3}{\cancel{2} \times 7} = \frac{6}{7}$

Part 12

$\angle$ t = 38°

Part 13

b. $\frac{11}{30} < .37$

Part 14

$\angle$ p = 81°

Part 15

24	42	66
88	11	99
112	53	165

66

Part 8

b. 1. 266 lb
2. 76 lb

d. 1. 75 students
2. 200 students

Part 9

b. C = 31.4 ft

Part 10

	x	y
B	3	9

Part 11

b. 1395 windows

67

Part 7

b. R

d. R

Part 8

	short	long	dogs
b.	$\frac{17}{66}$	$\frac{49}{66}$	$\frac{66}{66}$

d. 1. $\frac{6}{8}$
2. $\frac{2}{8}$

f. 1. 75 mph
2. 30 mph

h. $9\frac{1}{3}$ min

j. 26 doctors

Part 9

b. $\angle$ c = 38°

d. $\angle$ t = 129°
$\angle$ z = 51°

68

Part 5

b. 400 lb

d. 1. 75 hr
   2. 225 hr

Part 6

b. [270] ÷ 2 = [135] (? above 270)

[135] ÷ 3 = [45]

[45] + 50 = 95

Part 8

b. 1. 128 jackets
   2. 160 jackets

d. 264 mi

f. 108 lb

h. $97\frac{1}{4}$ mph

Part 9

b. C = 56.52 yd

d. r = 6.37 m

f. P = 80 ft

69

Part 6

b. 1. $418
   2. $22

d. 1. 40 miles per hour
   2. 16 miles per hour

Part 7

	x	y
B	3	8
D	6	7

Part 8

b. $9\frac{1}{2}$ qt

d. 52 men

f. 1. 52 men
   2. 39 women